A natural ecology

Michael Graham
CMG OBE(mil) MA

A natural ecology

SOMERVILLE COLLEGE
LIBRARY

Manchester University Press

Rowman and Littlefield
Totowa, New Jersey

© 1973 Manchester University Press
All Rights reserved

Published by
Manchester University Press
316–324 Oxford Road
Manchester M13 9NR

ISBN 0 7190 0483 7 (hard covers)

ISBN 0 7190 0529 9 (paper covers)

U.S.A.
Rowman and Littlefield, Inc.
81 Adams Drive
Totowa, N.J. 07512

ISBN 0 87471 027 3

16114

Designed by Max Nettleton

Made and printed in Great Britain by
Butler & Tanner Ltd, Frome and London

Contents

		page
	List of illustrations	vii
	Publishers' acknowledgements	x
	Preface	xi

1 *Levels of ecology*
The class—Haeckel's ecology or autecology—
synecology—macro-ecology—practical work   1

2 *The farm/I/inhabitants of the soil*
Farming and soil science—roots—the larger fauna:
earthworms, wireworms, etc—mites and smaller—methods
of investigation—some other micro-organisms—fungi—
actinomycetes—bacteria—clay, humus and chemicals   13

3 *The farm/II/the web of life in the soil*
Animals and grass—consumers of roots—mulch:
earthworms—acidity and oxidation balance, pH redox—
mineral efficiency: nitrogen—soil organisms in action—
cycles—arable husbandry   25

4 *The farm/III/the great debate*
Of chemical manures—history of controversy—the muck
and magic school—the controversy decided: main facts—
particular facts—even dustbowls   45

5 *Aquatic ecology*
Ponds—streams—lakes—estuaries—the rocky sea shore—
nearby seas—the ocean—general review—pollution   60

6 *Forestry*
Properties and uses of timber—ship timber—virgin forest—
plant succession and climax—Ladang—conservative
lumbering—trees for conservation—plantations—volume
of standing timber   83

7 *Woodland*
A book—habitats—niches—pyramids—cycles—air
pollution—general remarks   96

Contents

8 Migration of birds and fish
A part of ecology—tern and plover—land-marks—skymarks, migration fidgets—sun as guide—stars—spawning and feeding—Meek's Law—niche-filling—observant fish and birds—England and Africa   *page* 108

9 Predators and prey
The balance of nature—number control—therapy by selective predation: territory—emigration—parasitism—human reactions—major fluctuations   124

10 A tropical Great Lake
Speculative preparations—history—survey—fishing—drift and wind—fixed station—testing an hypothesis—solution—retrospect—the determinant—scientific research—awkward facts   141

11 Population numbers
Morand and Laplace—Hensen—Petersen—direct methods—Southern—improvements—Petersen's intuition of the optimum catch—fish husbandry   161

12 Erosion/I/general notes
Magnitude—rock to soil: animals—soil to rock: animals—Rivington—Kersal Moor—wind erosion—detachment of mat—more about trees—Savory on Rhodesia—mismanagement of animals   178

13 Erosion/II/swift rivers
History—observations in Borrowdale—observations in Kentmere: major discussion—tentative advice based on the theory   194

14 Reclamation of spoil lands/I/technical
Something immediate—practical trials—pilot scale—the Easter Course: notes—policy   207

15 Reclamation of spoil lands/II/social
Introducing children—recreation—frustrations and vandalism   219

16 Ecology in progress
The theory of determinants—a common ecology? Advice—scholasticism: logic on too few factors—clearing the decks   228

Bibliography   235
Index/names   243
Index/places   245
Index/subjects   247

List of Illustrations

Chapter 1
1 Autecology — page 4
2 Cod country — 5
3 Synecology — 6
4 Deeper synecology — 7

Chapter 2
5 Larger soil organisms — 16
6 Smaller soil organisms — 17
7 Soil bacteria — 17

Chapter 3
8 Cultivators — 28
9 PH range in which crops occur — 32
10 A faulty area of grazing — 35
11 Fertility cycle — 40

Chapter 4
12 Soil deficiencies — 47
13 Wind erosion on light soil — 58

Chapter 5
14 Synecology of ponds: plankton and pondweeds — 61
15 Synecology of ponds: the detritus — 62
16 Thermocline and seasons in a lake — 65
17 Seasons under a thermocline in the North Sea — 66
18 Limpet the great grazer — 69
19 Clarke's web diagram — 71
20 Tracks of cruises on the supply of nutrients in the North Sea — 72
21 The nutrient cruise in 1935 — 73
22 The nutrient cruise in 1936 — 74
23 Mortality of plaice in near waters — 75
24 Dominance of fishing in near waters — 76
25 Layers in the Atlantic Ocean — 77
26 Contention of waters as key to location of fish — 78
27 Contention in profile — 79

Chapter 6
28 Canopy makes proper timber — 84
29 Estimating timber height — 92

30	Opportunity from fire	page 94
31	Order from fire	94

Chapter 7

32	Elton's pyramid	102
33	Food chains	104
34	Air pollution	105

Chapter 8

35	The migration of Arctic terns	109
36	Drift of eels	114
37	Migrations patterns recorded on scales	115
38	Migration of plaice to the Southern Bight	118
39	Plaice migration from Dogger	119

Chapter 9

40	Leopold's concept	127
41	Morrow's unicorns	129
42	Kirkman's robins	130
43	Territorial behaviour in animals: social contexts	133
44	Territorial behaviour: interloper	133
45	Territorial behaviour: new claim succeeds	133
46	Territorial behaviour: a further claimant	133
47	Fluctuation in herring	139

Chapter 10

48	The fishing survey of Lake Victoria, 1927–8	142
49	The staple fish	142
50	Local fishing	143
51	Net fishing	144
52	Local fishing station	145
53	Scenery in ngege country	145
54	Scenery where ngege were uncommon	146
55	Exposed coast	146
56	Intermittent south-east trade wind	148
57	The fixed station	150
58	Victoria Nyanza, southern half: conditions	152
59	Southern half: ngege fishery	153
60	Older ngege	154
61	Escape of an awkward fact from the Press Censor's Office	159

Chapter 11

62	Census by eggs	162
63	Total mortality rate: Netherlands soles	168
64	Overfishing	169
65	The optimum catch	171
66	Whales	176

List of illustrations ix

Chapter 12
67	Erosion in the English Lake District	page 181
68	Classical depletion by erosion	184
69	Modern erosion: neo-pagans on Rivington Pike	185

Chapter 13
70	The Borrowdale Derwent	197
71	Tell-tale stump	197
72	Coombe Beck protected	198
73	Coombe Beck devastated	198
74	Coombe Beck mill	199
75	Coombe Beck: cavitation in sheep run	200
76	Some complications on banks	201
77	The method at Kentmere	202
78	Forces on river banks	203

Chapter 14
79	Natural regeneration	208
80	Spread of trial plots over South Lancashire	210
81	The right riders	211
82	Footprints	211
83	Sallow pegs	212
84	Growth of pegs	213
85	Slope of willows	213
86	Sallow patch	214
87	Terraces	214

Chapter 15
88	Young energy	220
89	Their rucks	222
90	Playsward on colliery shale	225
91	Scars	226

Chapter 16
92	Ancient lagomorph adapts	230

Publishers' acknowledgements

The death of Mr. Michael Graham having occurred very early in 1972, when his manuscript, though complete in all respects, had not gone into proof, his publishers are deeply indebted to the following individuals for their help and advice in the course of the book's production:

Mrs. Edna M. Woodrow
>who, having acted as Mr. Graham's assistant with the preparation of the typescript, took over the proof reading and compiled the Bibliography and Index;

Professor D. H. Valentine
Mr. Gordon Blower
Dr. D. H. Cushing
Dr. A. J. Lee
>for valuable specialist help;

Mr. Andrew Becker
>who redrew, often from rough sketches, many of the maps and diagrams.

We are also grateful to the authors and publishers of previously published illustrations for permission to reproduce them here.

Preface

I hope the young people of today will like ecology. If I understand them, they should do so; because ecology takes class-room subjects like physics and chemistry out into the open air and builds them into a new subject with facts already there. 'Thoughtful natural history', is Dr. M. J. Parr's definition of ecology; that is, firmly based on the facts, of natural history. H. N. Kluvers suggested a definition to Mr. Zweers of Amsterdam, who told it to me: 'the knowledge for attaining the balance of nature.' This seems more ambitious, but is none the worse for that. We could mention here what ecology is not: neither definition lets in speculation on the optimum level of human populations, because there are no usable facts on that extremely complex subject: none that could bring us nearer to an estimate. William Kershaw, Professor of Biology in the University of Salford, aimed also at adults, when he published his paper against blanket pesticides in the *Proceedings of the Society for Occupational Medicine*, 1966, showing a better technique based on ecology. In the same year he instituted the course of twelve lectures on which this book is based. I knew of no such subject as well-balanced as ecology for as few as twelve lectures. What a pity! I was sure that the professor was right, so could hardly avoid the task of making a course up. For authenticity, it came out as mainly from my own experience, supplemented by what I had seen something of, in nearly fifty years. Thus—my kind colleagues say—I broke new ground, and when putting it together for a book I encountered unities—for example, Pearsall's redox—which I offer as integral parts of the subject, to be neglected by my fellow ecologists at their peril. There are several more, such as the optimum catch.

M. J. Parr and P. St. J. Edwards read every word in draft; the Librarians of Salford University Library, the Fisheries Laboratory at Lowestoft and Chetham's Library Manchester and Dr. Frankland of the Nature Conservancy Station, Grange-over-Sands, provided valuable help. I have to thank all these and others too numerous to name.

We may not, after all, have found the essence of ecology, but we may go some way on the journey if we start right, and begin with Haeckel, as I shall attempt to do in Chapter 1.

Levels of ecology

The class

This book originated in a course of lectures. The Royal College of Advanced Technology in Salford was about to be reclassed as a University. It had recently acquired a full professor in biology, William Kershaw, and in 1966 one of his innovations was a course in ecology, the content of which he left entirely to the lecturer. It was for the third-year students of an honours degree course in applied biology, and there was to be just one term of twelve lectures, with the accompanying practical periods. This book is based on these lectures, but most of the studies are expanded and a general review added as Chapter 16.

The course formed what seemed an opportunity that might well be unique in the students' lives, and I thought that it would be wasted if we spent all of that short time collecting animals or plants and naming them and mapping them and so on, in all the detail that is usually associated with ecology. Surely in these days when planning and landscaping are of great topical interest, one could do something in a more general way that would yet have precise usefulness.

Many years ago I had been very impressed by seeing at the Rothamsted Experimental Station a corner of a field that had been fenced off and which over the course of years had turned into a flourishing woodland. It seemed to me that was a case of behaviour of landscape. It did not appear to need a detailed knowledge of what grasses were suppressed in the field and what trees arose. That example seemed to illustrate the sort of ecology that I should like to aim at, in order to give this class of students with limited time some understanding and awareness of the properties of landscapes of various kinds.

Haeckel's ecology or autecology

The term ecology was invented by Haeckel who might be said to have invented ecology because he invented the term. John Ruskin always referred us to the derivation of terms, saying that often important aspects of the originator's thoughts become lost by the way in later usage. This one seems to mean knowledge of the housekeeping.

Haeckel's conception was of the animal species and all its connections to the environment, its habitat: in the case of a fish species, the kind of water it lived in, its food, what eats it and what competes with it. Among other things his conception formed the basis of fisheries science, which arose in the following way.

In 1883 there was a great international conference on fisheries in which practical men—that is, fishing skippers who had become owners of smacks and had kept records over the years of their annual catch—showed that annual catches per smack of such valuable species as sole and turbot had fallen year by year. They considered that to be a major problem, which had to be solved; it showed an unhealthy state, even though they themselves had made good money.

Biologists responded by offering Haeckel's ecology and practising it. In the fullness of time in 1920, I was instructed to study the ecology of the cod species and to use all the facilities available to find out its *autecology*. That is a technical word, and not a particularly euphonious one, but it is better to use it perhaps than to say 'Haeckel's type of ecology' or 'specific ecology' or 'particular ecology'. The word autecology means Haeckel's ecology as it has just been explained.

In boyhood I had been taught to take an interest in animal behaviour in a statistical way when I was too young even to have heard of the word average. My elder brother and I were watching some tufted duck diving in Esthwaite Lake. I was interested in discovering the maximum time the duck stayed submerged, but my brother explained in a lofty manner that the average would be much more significant and so we would work out the average. This seemed to me to be a very fine thing to do at the time and I have never quite lost the taste for it.

Consequently, I was very pleased when in my task on the cod species, I was handed many sheets of fishery statistics. They were of two kinds, some were of hundredweights of fish collected by the Ministry's men on the fish markets at the ports, together with the fishing grounds from which they came and the length of time that the ship had been at sea. The figures for the trade category 'large cod' yielded very good results in a short time, namely the spawning grounds and spawning season and the annual cycle of migrations for feeding in the North Sea. In order to produce those results I needed only the minimum of supplementary biological investigation, and by beginning with a love for vast statistics the answers were necessarily in proportion from the start.

Another set of sheets was of the body-lengths of fish specially measured by observers who went to sea in commercial trawlers. After three months of trying to extract some intelligent data from them, I had to put them away disconsolately, and only after fourteen years of other research did I have the confidence to mass all the data together by months and make out a very good statement of the growth rate of the cod. I mention this chiefly to show that Haeckel's ecology, on which all the rest is in practice based, involves a great deal of drudgery; but I am glad to say that it was not all paper work in the office.

Reaching down to the bed of the sea we used grabs and dredges for the animals that form the food of cod, and we identified and counted them. There were cod stomachs to examine and thousands upon thousands of scales and earstones, in order to determine age. The predators of each species of fish were traced, then there were calculations of growth rates and mortality rates and the rate at which the stocks were being fished. In the end one could say 'fish more' or 'fish less', but not from study of autecology alone.

The present discussion is about autecology and I think that the history of fisheries disputes will serve to show how essential autecology is. Some said fish more, some said fish less, and so the controversy has gone on all over the world, for rivers, lakes and seas. The human mind naturally asks for more knowledge bearing on the problem, and that is why the biologists answered the skippers and owners by starting to practise autecology.

We have said enough to have seen the sort of methods used in fishery science and it will be useful I think, by way of illustration of the general subject, to specify rather more. I am taking the results of study of the cod, to which I was a devotee to the point of obsession for a decade from 1920–30. I do not suggest that everyone should know the details of the autecology of the cod, but only that they should know what is involved in autecology, and the cod will do for illustration as well as any other species.

First we may look at the late Dr. A. V. Taaning's chart, which I have extended (Fig. 1). It shows the tracks of wide-ranging marked cod in the north Atlantic and a very fine extent is shown, ranging from near Nova Zembla to the coast of Virginia in North America. No one cod has been recorded as going all the way, but some wanderers have travelled from one cod region to another jointly, to cover the whole range, so that we can be fairly certain that any dearth of cod stock in one region would soon be filled up by specimens migrating from another. That is, there can be no

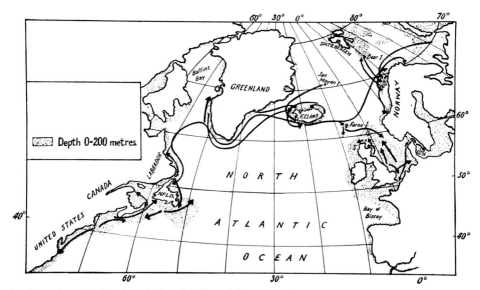

1 Autecology. Tracks of marked cod (collected from 14 authorities) show that this species is taken on banks in the Atlantic near the regions of ice. Adapted from Taaning (1937) and reproduced from Graham (1956)

question of fishing any cod right out so long as there are reserves elsewhere. Whether there still are great reserves in regions more remote than the North Sea is another question, but it is highly unlikely that any of the remote stocks would be fished so hard that there would not be a few wanderers to fill any gaps left by disasters in one area or another. The fishery would cease to pay long before then I am sure, so we may take it that is the meaning of Taaning's chart. Having said that, it must not be concluded that cod do in practice wander about so far. The evidence is that each well-known stock, such as those of the North Sea and other seas bordering the British Isles, even those so close to each other as those of the Faroe Islands and of the Faroe Bank, do usually stay where they are and have done so for long enough to be genetically distinguishable from each other statistically on kinds of proteins, as Dr. Jamieson has found at Lowestoft.

The cod does not confine itself to one kind of food. It eats almost any kind of fish, but the most important of its foods is the herring, which is usually to be found in upper and middle waters. Many of the other species on which the cod feeds live on the bed of the sea or even burrow into it a little way. The cod has a well-developed barbel under its lower jaw, which is thought to be connected with this habit of hunting about on the bed of the sea, and its enormous gape, which opens its whole face when it opens its mouth, is thought to be connected with feeding on actively swimming fish in the open water. With the evidence from the stomachs showing such flexibility, it is not in the least surprising that cod is such a widespread and successful species.

When we come to consider what the herring feeds on, or even what the haddock feeds on—for this is an animal of the sea-bed also taken by the cod in significant numbers—we shall step away from the autecology of the cod and consider synecology, or the web of life.

Synecology

Synecology is a word that reached me from my colleague Dr. Popham, spelt differently from my way. He describes it in terms of himself and his young pupils in his book *Life in Freshwater*.

2 **Cod country.** The mate of the Arctic Research vessel *Ernest Holt* has called the crew and has climbed onto the fo'c'sle head to start knocking ice off the upper works where it has formed out of the frosty air, its weight forming a hazard to the ship's stability. (Photo, N. Reynolds.)

The study of the synecology of one habitat may be undertaken by a group of workers, each of whom studies the autecology of a few closely related species. Group discussions are held at intervals when each member of the team gives an account of his own observations of the animals he is studying. From this information contributed by every member of the team, a picture of the whole of the habitat can be synthesized. Every member of the team sees the significance of his own investigations in relation to those of other workers. This approach never fails to stimulate interest and demonstrates the complexity of the web of life, even in a small habitat.

Dr. Popham speaks of a small habitat, but his description will do just as well for the work of a sea-fisheries laboratory, connected as it is with research ships working perhaps from Greenland to Nova Zembla, from Malta to Newfoundland (Fig. 2).

Having brought the cod into relationship with the herring, we find that the web of synecology extends right down to unicellular plants such as the diatoms and flagellates drawn by Sir Alister Hardy (Fig. 3). Hardy is surely our most artistic Professor of Zoology and he originally made this diagram largely from Marie Lebour's work, but also including his own observations and those

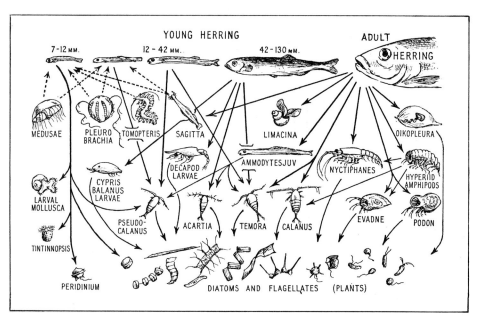

3 **Synecology.** Especially through its favourite food, the herring, the cod is based on plankton, down to the smallest microscopic plants. Arrows show the direction of predation. From Hardy, Savage, Lebour, *The Open Sea* (1959), by kind permission of Sir Alister Hardy and Messrs. Collins, London.

of his colleague at Lowestoft, R. E. Savage. As I sketch this diagram on the board in the lecture room, I add the cod which feeds on herring, and feeding on the cod there would be seals, and feeding on the seals there would be polar bear and Eskimo. There would also be the numerous fishermen who in historic times have fished

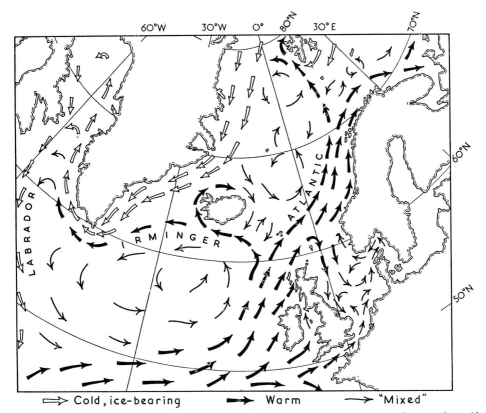

4 *Deeper synecology.* Lee's chart from Graham (1956), which would continue on the same theme if carried to the westward to cover the whole cod area. The theme is part of Wimpenny's orderly turmoil (1966) (see also Fig. 27 in Chapter 5 below). The disturbances visible here as currents on the flat, affect the deeper layers. They are liable to variation from wind and storm, as consequently are all the related effects, but the system is frequent enough to be worth describing and to be ecologically effective: that is as 'determinants'. Banks and ice (Fig. 2), each have disturbing effects, but essentially, the system brings water from dark situations to the well-lit layers where alone plants can grow. The mixing should not be so much that they have not time to grow, as Gran and Braarud (1932) found at the mouth of the Bay of Fundy—that is not usual. World wide, water from dark depths has accumulations of nutrients, N.P.K. etc., because it has had no plant growth, and the same is true of water from under ice. When this water is brought to the well-lit layers, the system of Fig. 3 can begin, culminating in the herring food for cod; and also, in the rain of dead animal plankton for the sea-bed, feeding bottom-living animals, on which the cod and other fish also feed. Thus came about the great fisheries. But to judge their changes and predict their future, their *macro-ecology*, we need statistical and historical evidence, as well as the foregoing. Failing that, pilot-scale experiments are called for.

the cod, especially in the better known grounds such as Newfoundland, Iceland and Norway. Competing with the cod for herring we should have to add gannets and toothed whales and competing with the seals for cod we should add Greenland sharks. Competing with herring for the krill—animals such as the shrimp-like nyctiphanes which is drawn here—are the whalebone whales and many pelagic birds. It is evident that the more one knows about any eco-system, the more difficult it is to show a tidy diagram of its synecology, and that is true of the web of life as it exists in nature.

So much for the web of animals and plants based on the plankton and involving notably the cod. Part of synecology however, relates to the conditions of the environment and here I would show a chart drawn for me by Mr. A. J. Lee (Fig. 4). It does not reach to Newfoundland and the American coast as Taaning's did, for the simple reason that map projection would have made the shape of the currents rather unreal if he had depicted on a flat piece of paper so large a slice of what is really the curved surface of the globe. It will be remembered, however, that the famous cod area of Newfoundland is notable for great ice-bergs coming south from the Davis Strait, which are dangerous in the common fog. The melting bergs and fog are symptoms of the mixing of really cold and rather warm water. For the rest of the cod area we may proceed by comparing Taaning's diagram (Fig. 1) with Lee's (Fig. 4). We may first note that on Taaning's chart the water of less depth than two hundred metres is dotted in. Fishing methods have all along been developed in not much greater depth than that. In recent times fishing has reached down deeper and has also been prosecuted in middle waters of greater depth, but the famous fishing areas did depend on the fairly shallow water that is dotted on Taaning's chart. Now going to Lee's chart, we see that the area of the fishing grounds off Iceland is characterized by mixed cold and warm water. So are the grounds of the North Sea, the Faroe region, the Spitzbergen shelf where Bear Island lies between Spitzbergen and Norway, and doubtless those smaller areas which cannot be shown. This point might have been introduced under the cod's autecology but it is economical to have left it till we came to synecology, because the connection of the mixed water and the cod is via the plankton and herring. As well as the mixing on the surface, deeper water is pushed upwards by the continental shelf shown by the dotting in Fig. 1.

This water, having been in the dark, has not supported any active plant life and consequently is rich in nutrient salts, phosphates and nitrates, from the excretion of animals, whether within it, or sinking to it from above. When the deep water comes to the surface it meets many spores of microscopic plants liberated from the ice and so there is an active outbreak of the microscopic plants called collectively phytoplankton. Feeding in them at first at the fringes of the up-welling are to be found vast numbers of microscopic animals; then we find the pelagic animals, such as herring, that feed upon the microscopic animals; then the cod, feeding upon the herring.

That this sort of knowledge of synecology does not do much to answer the question by which the smacksmen in 1883 set the whole programme going, namely whether to fish more or to fish less, illustrates the limitations of ecological evidence generally. It is interesting, however, and has in the past been of great value in preventing short cuts leading to unjustifiable actions which may have seemed more than right at first appearance. Those who regard their local stock of fish as entirely self-contained are inclined to give it much more protection than is necessary. There are even ideas about fertilizing the whole sea by nitrate derived from the nitrogen of the air. Revelation of the vast stores of nutrient salts, including nitrate, in the deeper waters of the oceans makes such efforts seem puny. Nor does knowledge of the immense part played by nitrogen compounds in ecology really support the idea. In well chosen circumstances—laboratory, greenhouse, spring grass, sea loch—the action of nitrate may be simple and beneficial, through direct plant nourishment, but in the variation of real life there are complex effects and inter-actions.

Macro-ecology
Now to introduce a third term, macro-ecology. To illustrate it my mind turns back to that haunting example of the piece of field at Rothamsted that turned into a wood. Many people have realized that the fact was significant, without any of us having information on the species of grass or trees. We knew nothing of the autecology or of the synecology yet there was a piece of the subject which could not be called anything else but ecology, and it seemed to me that for this short course the changing field instanced the kind of ecology that should occupy the students. Later I came to the name *macro-ecology*—ecology of large units. I am told that this name is already used by some German ecologists.

It may be believed possible to find out and teach macro-ecology, but it is quite clear from what has gone before that we should be very unwise in practising it without having studied the necessary autecology and synecology. For example, we do not know whether the field at Rothamsted was rabbit-fenced nor whether this might have been an important factor. On light soil the starting of trees might have been greatly delayed by rabbits: the trees could hardly survive until there was a thick protective cover of brambles. Consequently, a man who was hoping to do the same thing deliberately would have to understand the autecology of the rabbit and of the little trees. But the bare fact at Rothamsted can be taught and, so far as teaching can go, it makes sense. To gain from the study of fields or woods it is necessary to know that fact of macro-ecology: the liability of the one to turn into the other.

It is as if the field were alive as one whole organism, or as if the resulting wood were alive as one whole organism. We may indeed in the same way think of any eco-system as a whole, live organism. A pond is alive, a canal is alive, an ocean is alive. We can learn what changes can be described in each eco-system, what responses each will give to different stimuli. This is not a far-fetched way of talking, for any organism, a fish for example, is a community of cells, as it were, an eco-system. It is rather more disciplined, it is true, than the eco-systems of ecology, but even in the body of a fish there are cells, some blood cells for example, that seem to act independently. Yet, in spite of that, the whole fish will very often react quite as simply as a single-celled organism will.

Thus we have arrived at the third level of ecology. The autecology of many species put together forms synecology and the synecology acting as one whole forms macro-ecology. These three levels are the subject of this chapter, to be used wherever they can be throughout the book.

It seemed possible that the syllabus could be approached in the following way. The field at Rothamsted turned into a wood because it was fenced. We may say that one factor was eliminated which would have stopped the change, namely keeping the herbage short by grazing. So we might say that a field or sward is determined by, it may be, grazing; when we take away that factor the eco-system changes. Shortness of herbage, in that instance, could be called the determinant.

The aim of the book might well be to see whether there are determinants for other eco-systems that can be stated so simply. It seems most improbable that enough will be known to state them dogmatically, but an enquiry on those lines would prove a fruitful course in ecological knowledge and be of considerable interest both in college and outside, especially where people are trying to improve landscape.

There is one fine example of the animal proving itself to be the determinant and performing a macro-ecological feat, namely the *Coypu*, or *Nutria,* invading the Norfolk Broads and clearing the former water-ways. The area was going back to the fairly dry land that it had been converted from by medieval peat cutting, silting up over the centuries having at last brought about a critical shallowness, so the vegetation could root easily and cause more rapid silting up, and start bushes and trees in succession. In 1937 a few Coypu escaped from fur farms and multiplied and spread quickly in a habitat that was very suitable as they fed on roots and stems of waterplants. Although naturalists deplore the consequent elimination of some species, the Coypu have restored the area to pleasure-boating, which was becoming restricted rather rapidly. This is described by E. A. Ellis in *The Broads.*

The main University buildings at Salford are sited on the banks of the River Irwell, 'the winding torrent', and there existed a wish or a plan to make more attractive a part extending from Manchester to Bolton along the valley of that once beautiful and interesting river which has become merely useful and doubtfully interesting but no longer beautiful. The North West Economic Planning Council had produced a very good report from a committee under Sir William Mather with numerous suggestions for improving the landscape, and we could feel that we were a part of that movement. Many people surely seek for a reasonable landscape controlled by moderate and understanding people. Towards that end it will be necessary to trace the determinants of each kind of landscape or eco-system. Thus came the syllabus of the lectures and the contents of this book.

Practical work

Several times in my life as a practical ecologist I have found myself called upon to make a sketch map without any measuring instruments. All ecological data tend towards making such maps.

We chose Kersal Moor which is part of the steep bank marking the edge of the River Irwell flood plain on the north side and Castle Hill which overlooks the valley. Having made our sketch maps, we entered on them the results of taking soil and samples and determining the moisture content and the acidity (pH). This might contribute a little to the project of the linear park, and the ground lies within the city boundary of Salford and within view of the college.

We assumed that we should not have instruments. The students determined their own body measurements and practised pacing.

We used five-foot garden canes for sighting and levelling, and plant labels to mark the ground.

The teaching followed that given by the late Professor Frank Debenham in his book *Map Making*. We had, throughout, the permission and kind interest of the Salford Parks Department.

The practical work was therefore rather detached from the course of lectures, in which, after introductory remarks somewhat as have been given in this chapter, we turned to the only subject that was foreseen and pre-ordained, namely the farm.

The farm/I/inhabitants of the soil 2

Farming and soil science

Practical farming is very difficult, and I would not venture to advise on it. But farm ecology is fascinating and deserves to be part of the riches of a well-informed mind.

When we see the rivers run down there is no harm in knowing that when the water runs down towards the sea some of it will be evaporated and come back in clouds, that there will be evaporation going on from the wet fields, and that when the clouds meet cold air they will drop the rain. The rain then goes on the land and into the land drainage—provided that there is some drainage. In nearly all of our land there is some drainage by which man has made the water leach more quickly into the little streams and so into the rivers again.

Knowing so much does not in any way reduce the beauty of clouds as John Ruskin illustrated it, and neither should knowledge of the ecology make the farm less attractive.

The late Professor W. H. Pearsall, one time President of the British Ecological Association, was the first well known man to apply facts in ecology collected for their own sake to the practice of agriculture. The application of his research on moorlands and meres was to the growing of rice, which is the staple food for more than half of the inhabitants of the world. Other people might argue with a good deal of justice that Sir George Stapledon, although he studied agriculture itself, directly applied the ecological outlook. So, although this chapter may be unique in following the ecological scheme that has been sketched in Chapter 1, it is all the better founded in that it follows the lead of two well known and original ecologists, Stapledon and Pearsall.

The autecology of soil and its organisms, with which this chapter will deal, has been learnt or checked particularly from the work of the Russells, the father Sir E. John, and the son Dr. E. Walter. The father produced the paramount book of the Rothamsted Experimental Station, *Soil Conditions and Plant Growth,* and his seventh edition was a model for his times, but was not nearly sufficiently taken to heart.

In constructing the present book, I used the ninth edition of 1961, in its fifth impression of 1968. Dr. Walter Russell rewrote the whole book in 1950 and further revised many parts in 1961. He gave a wealth of reference but did not lose the balance and good sense of the original work of the father. There is something attractive about an immortal book being tended by two generations in one family. It was one of the services of the father to introduce in the seventh edition such biology as was then known. Now in the ninth edition, it looks as if biology out-rivals chemistry, which has hitherto constituted most of agricultural science. It may not be long before biology is generally accepted to be a much more important component of agricultural science than is chemistry. The Russells' title assumes that soil conditions in general do govern plant growth, but that was not always realized: there was commonly a restricted reference to chemical salts, to which we shall come in due course. Let us try to do a little better starting from another beginning. If soil conditions act, then it must be through roots.

Roots

Certain simple kinds of plants, such as algae, lichens or mosses, can live on stones or on other surfaces; but plants growing on soil have a better store of moisture, which they tap by roots. The soil is something like a sponge, in which roots can freely ramify and obtain energy by breathing the air in the spaces and absorbing moisture for the use of the plant. A sponge has what is called a lattice structure, and so has the soil.

Providing that the soil is not waterlogged but is well-drained, the plants put down new roots very quickly, replacing rootlets that become obsolete. Only about one-tenth of the length of a rootlet is absorbent; the rest is a pipe for conducting water and other things to the main body of the plant, for which purpose it is rendered water-tight and air-tight by cork. The cells in the surface of a rootlet tip are extended into very fine hairs, known as root hairs, which attach themselves to soil particles and absorb the soil solution or part of it. Soil solution is mainly in the form of very thin coatings of water on soil particles, on vegetable remains and on some of the micro-organisms. It is of importance, long-known but only recently appreciated, that roots of water and marsh plants have the capacity to bring oxygen down to the mud via air-spaces in their cells, so that the oxygen can pass down from the part of the plant that is up in the air.

The larger fauna: earthworms, wireworms, etc.

Moles and voles, slow-worms, snakes, toads and slugs all inhabit the soil, but their weight per acre is very small. That observation shows the need for some idea of the relative quantities of organisms in the soil. An acre with a hundred mole-hills showing looks heavily infested, but actually there are probably only but two moles, at 1 lb weight apiece, whereas the cattle may be half a ton per acre, and the earthworms nearly as much. Those weights are called biomass. The earthworms have been estimated at Rothamsted to number up to a million per acre of deep rich soil but that may be an over-estimate. It would mean two hundred per square yard, and few gardeners will have seen anything like that number when digging. The discrepancy may be accounted for, partly, by young ones, by small species and more particularly by the scientist's estimates going much deeper than the digging spade. Nevertheless, it may be better to think in terms of a fraction of a million rather than of a full million of earthworms, per acre, in most soils.

One of the components that is found fairly frequently in soil is the group of animals known as myriapods which are centipedes and millipedes. The centipedes feed on earthworms and other animals, and the millipedes on vegetable matter. The latter are often called wireworms but this name is perhaps better kept for click-beetle larvae, which are also called wireworms by the scientists. Both groups assist in their way in cultivating the underground soil. Millipedes make a little capsule of soil in which they lay their eggs, and thereby, if there are enough of them, they make a little crumb-like structure. The centipedes do something similar. The female centipede liberates a rather big egg, and then, in order to prevent the male from seizing it from her and eating it, she plasters it over with soil, making a similar capsule or crumb. Many kinds of beetles and their larvae are active in the soil; also flies and ants.

In grassland true wireworms (Fig. 5b) are much more abundant than the myriapods. These click-beetle larvae feed on the constantly dying grass roots and can also attack living roots, as do many other beetle larvae.

Mites and smaller

The next heavy group, is that of the mites (Fig. 5c), which are very small members of the spider order. Some can be seen only with the help of a microscope. The familiar little red spiders that we see on birds, and sometimes about the place, are in fact some of the larger

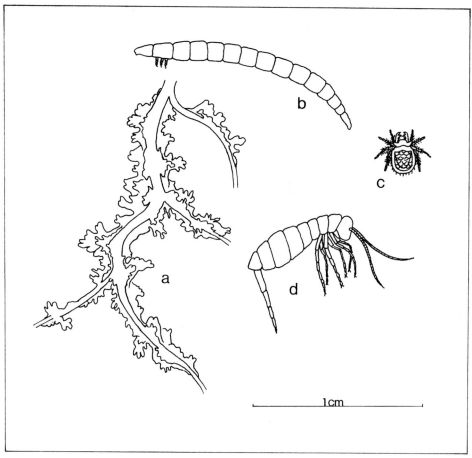

5 **Larger soil organisms.** The weight of live things in fertile soil amounts to a few tons per acre, from earthworms down to bacteria; and the bacteria usually weigh as much as the worms, which in a good pasture amounts to that of the cattle above, say half a ton per acre. Some samples shown here are (*a*) rootlet bearing mycorrhiza (fungus), (*b*) wireworm (beetle grub), (*c*) mite (arachnid) and (*d*) springtail (primitive insect). Drawn from the Russell's *Soil Conditions* and Sharp's *Insects*.

members of the mite group. Mites exist in the soil in millions per acre and collectively have quite a considerable weight, in spite of their small size. They include species that can live on simple vegetable remains and others that suck blood or sap and are entirely parasitic. Some of them are moderately versatile and it seems clear that many of them manage to digest cellulose, which is always a problem for larger animals, even herbivorous ones. Mites are the principal consumers of the hair-like hyphae, as they are called, which form the subterranean part of the mushroom type of fungi and of the moulds of the soil. As a fungal hypha dies back behind the active tip, the mites clean it up. It seems that the fungi unaided can consume even lignin, which is the toughening substance in wood, and so convert it to the tissue of their own hypha walls, which the mites can eat eventually.

The mites render two special services to the physical conditioning of the soil: they make webs, as do their larger relatives the spiders, and these hold particles of soil together. The second service is that some species of mite have the habit of covering their backs with soil and scuttling about, thereby doing a certain amount of underground tillage.

Another very numerous group of soil organisms is formed by the springtails (Fig. 5*d*) which are rather simple and primitive flightless insects, capable of jumping in a random way, using the appendages at their hinder end, which are folded under the body and held on a catch, except when they want to jump. They exist in numbers comparable with the mites and eat various vegetable remains, but so far as we know they do not have the variety of feeding habits that we know of in the mites.

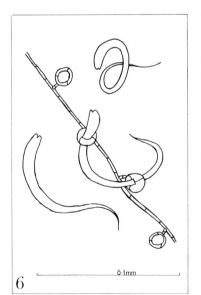
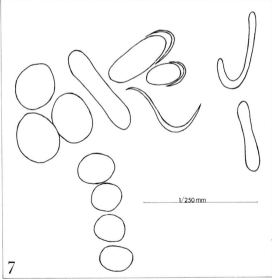

6 Smaller soil organisms. Nematode worms are abundant in soil, and some moulds have developed special snares to capture and hold them while they are digested by branches which quickly grow to penetrate them. Reconstructed from figures in Maio (1958). *7 Soil bacteria.* There are many kinds of these, including some that have flagellae (whips) which enable them to swim. In brief, the organic world, illustrated by Figs 5–7, is nourished by plant residues, and in its wastes coat clay and sand, making fertile soil.

The species of roundworms, or eel-worms, are called nematodes (Fig. 6). The group includes some very large worms, like *Ascaris* which is parasitic on the horse, but those in the soil are very small indeed, a millimetre being a considerable length for a soil eel-worm, although a few are larger than that. They are exceedingly numerous, often two millions to twenty millions per square metre, giving a weight of fifty pounds to one hundred and eighty pounds per acre. They are active in soil and some kinds do severe damage to plant life, whereas some may feed on other eel-worms. The part they play in the life in the soil has not been determined, but, considering the large numbers, it can hardly be negligible. They have some special enemies, of which perhaps the most noticeable are certain coiled fungi that form traps which digest nematodes (Fig. 5a). There is also a group of bristly little relatives of the earthworm that feed on nematodes. There are up to ten thousand of these bristle worms per square metre.

Although the part played by eel-worms is uncertain they probably consume some of the single-celled animals in the soil, which are called protozoa. These can be quite numerous and are of three main kinds. The amoeba-type creeps around in its well known manner and has a rather shapeless body; a variant has a shell, for example there is one called *Difflugia* which is a little more complicated than the naked amoeba, and there are the types which have a flagellum by which they can whip their way about through the soil solution. The third type is the hairy or 'ciliate' and this group includes a variety of forms, all predatory.

Most of these protozoa feed on bacteria (Fig. 7). Some of the flagellates such as the *Euglena* have chlorophyll and can make their own food from carbon dioxide in the air like a plant, although they whip or squirm their way about like an animal.

Methods of investigation

It might at this point be of interest to know how knowledge is obtained about these very small organisms. Soil scientists have found some ingenious ways of seeing them. They can place a few grains of soil on a microscope slide and put it in a cool oven for a few days and note what develops round the grains of soil. They can make a cell of soil about one millimetre in diameter and one millimetre deep with a hole in the middle and after a few days watch what has grown out into the clear middle. By the use of new staining methods they can tell live or recently dead tissue from tissue that has been dead for some time. In that way they can distinguish such small and pale things as protozoa and bacteria when on the soil particles or the humus particles, to which they are often very closely applied. Other ways include the culturing and counting of bacteriology. By use of the electron microscope of these times, they are able to obtain much greater magnifications of organisms which were too small for them to be seen in detail through former instruments.

Some other micro-organisms
As well as the familiar single-cell protozoa there are some that seem to be amalgamated into large multinucleate forms which are undoubtedly of some importance in the soil. They, like the protozoa, feed mainly on bacteria, which need a special section. But before going on to the bacteria we must notice the plants; first the algae-like *Pleurococcus* of walls and tree trunks. These are also found in the soil, and probably have been there from time immemorial, dating it seems from before soil even existed. The green ones are near the surface, and undoubtedly the algae include some kinds that can fix nitrogen as well as carbon. There are other plant groups; such as the diatoms which have silica shells; and especially the blue-green algae which do not need as much aeration as the green algae, but some can fix nitrogen of the air certainly, and probably carry out other chemistry.

At this point myxosporidia may be mentioned. These are inconspicuous micro-organisms forming spores as fungi do, and are probably especially important in woodland soils.

Fungi
Undoubtedly the largest and most important group of plants in the soil are the fungi (Fig. 5a). These have already been mentioned as food for mites. Here we are dealing with organisms that are usually associated in our minds with the breakdown of complex flesh or carrion or vegetable remains, rather than living off the simple carbon dioxide or nitrogen, or both, in the air. Thus fungi are characteristically saprophytic, that is, scavenging. There is little doubt that the first organisms to attack sizeable plant remains are fungi, and provided that they have some air they will dominate the decay situation to begin with. In their conditions of nourishment they range from feeding on complicated molecules, as in the carrion mentioned, down through the simpler vegetable remains. They include many parasites which attack living animals and plants and are responsible for a great many diseases from which crops suffer. There are members of the group which make the nut-like crumbs that we see when a fertile soil dries. Fungal hyphae have

penetrating power, and are liable to enter into almost every chemical reaction that one may meet with in soil.

A particular group of fungi is formed by the yeasts, some of which may fix nitrogen of the air and some of which can live without a very plentiful supply of oxygen, that is, they are anaerobic as opposed to aerobic. However, most of the fungi are on the aerobic side. An activity of fungi of great interest is the forming of mycorrhiza which are found inhabiting the roots of trees and other plants. They are characteristic of healthy stands of trees on poor soil.

There can be no doubt that the soil fungi as a whole are situated in the main stream of the nutrition arrangements in the soil, breaking down a great deal of old vegetable matter, eventually to make it available for new plant growth.

Actinomycetes

Actinomycetes form a section of bacteria, which means that they are more versatile even than fungi. They can feed on a great range of proteins of different complexities. They occur in the form of fine hyphae, giving them something of the penetrating power of fungi. The hyphae easily break up to form spores, which can be blown about widely. With such qualities it is evident that actinomycetes are inhabitants of the soil that are by no means to be overlooked. For well or ill they are powerful.

Bacteria

We come now to bacteria, mentioned several times already, of which there is a good deal to write. Bacteria (Fig. 7) are very small, about a million could be packed on one pinhead, and they exist in half a dozen well recognized forms. Their nutrition depends on a dozen or more levels of complexity in their food. Some fix atmospheric nitrogen and thereby build up substance, whereas others seem able to live only on advanced protein such as that found in living animals or in carrion. It is not unusual to find a scientist dealing with some forty or fifty bacterial species, so-called, but there are several reasons why they cannot be given the same evaluation as the species of higher organisms. For one thing they breed and vary too quickly. Furthermore, a strain that is flourishing on one foodstuff can be trained, as it were, to subsist on another. All this makes for rather difficult nomenclature, because, since bacteria reproduce so very quickly it is not the individuals that are 'trained' to do these things, but the strain gives rise to types more suitable to the changed conditions. Some form spores; some bacteria can swim. On the whole the main feature of bacteria is this extraordinary versatility. Having said that, we must nevertheless return to the idea that they have some specific quality, in order to describe what they do; but remembering always that these qualities are not fixed but are perpetually evolving. Of the higher organisms, by comparison, it is I think true that even they can only be distinguished into species easily, and in a rigid way, in those groups that have virtually stopped evolving. Such a stoppage has not taken place in the bacteria and it is difficult to believe that it ever will.

There are also, by the way, some organisms smaller even than bacteria and parasitic upon them. These are called phages, and though alive consist merely of a small cluster of molecules. Among bacteria, we have mentioned those which can fix nitrogen and build up protein from that simple gas. Another group of bacteria can dispose of sulphur provided that there is some nitrate present, and lime. Lime does many things for fertility of soil and it is accidental that sulphur disposal has been mentioned first. The importance of lime can be shown by mention here of only two more details: causing clay to flocculate and nourishing earthworms. The bacterial reaction for disposal of arsenic is probably very much the same as for sulphur; were it not so, the soil of orchards would surely be poisoned by the accumulation of residues of arsenical sprays.

There are bacteria that will degrade fuel oil when it is floating on the sea, so that it forms heavier lumps, which sink. In the soil, besides the nitrogen fixers, we know of bacteria that oxidize ammonia to nitrogen, nitrite to nitrate, and bacteria that will take the actions in the reverse direction. Thus aerobic and anaerobic bacteria are involved in the decay of organic material. The two types are used in the old fashioned method of sewage disposal, whereby the crude sewage goes into a tank shut off from the air, where the organic matter is entirely liquified but still stinks. The stinking liquor is then exposed to the air with the sprinklers going round, and the aerobic bacteria then take over the sweetening of the liquid, using the oxygen of the air. At which point, old fashioned operatives would demonstrate that you can drink the resulting clear solution, giving the comforting impression of being completely purified and harmless, which alas is not so.

Those bacteria that make colour pigments, and those that are fluorescent, do not seem to be eaten by the protozoa or by other bacteria, and those which can excrete antibiotics are not approached by predators either. Otherwise, as a group, they form the staple food of many of the micro-organisms already mentioned.

The importance of bacteria, in spite of their small individual size, can be seen when one takes the estimates of the total quantity. The weight of bacteria under the soil is of the same magnitude as that of the earthworms. I have tried some calculations in order to visualize what this means and I can put it this way—if the surface of the soil were bread and the bacteria were butter, there would be enough butter for a very mean scrape all over the field. That would not look like a very great weight but it does mean that there is an appreciable quantity of bacterial flesh, just as there would be an appreciable and just detectable spread of butter, but the bacteria are not spread on the surface. Another comparison gives the order of magnitude of beast's flesh on a sward as the same as that of the worms below it, and the bacteria would add an equal contribution.

It is of considerable importance to know where bacteria are really found in the soil. A great many of them are closely applied to the root hairs of plants, so much so that the word *rhizosphere*, or 'world of the roots' has been coined. Consequently the importance of the health of the community of bacteria to the health of the plant can fairly be surmised. Also, attention should evidently be paid to roots, more attention perhaps than they usually receive.

In general, in aerobic conditions, the bacteria excrete what the plants need, but in anaerobic conditions the bacteria are liable to excrete toxic substances, including stinking amines such as cadaverine and putrescine. An exception is in the particular rhizosphere associated with aquatic plants. There, as already noticed, oxygen is obtained from the rootlets themselves, reaching them via the airspaces. Thus the bacteria, even though they may be surrounded by anaerobic mud, as in the case with rice, yet can produce substances which appear to do the rice no harm but to nourish it.

Clay, humus and chemicals

The arrangements in aquatic soil seem all the more remarkable when we reflect on those in terrestrial soil, which, as we saw, is in the form of a lattice. There, plant production uses not only the soil cavities with air in them and wet layers on the surface of the lattice, but the substance of the lattice itself, in so far as it is partly composed of clay and humus, or, even clay-with-humus. Both clay and humus have intrinsically the capacity to swell up and absorb water. This water is only loosely bound and can be extracted from the clay and humus even in apparently dry soil. It is not quite correct to talk of clay and humus as separate entities except in the early stages of humus breakdown when there are still solid and even recognizable remains of plant tissues in the substance of it. That is humus according to the common usage of the word and not as defined by some of the men in the laboratories. Admittedly, it is not possible to treat humus in the way that a chemist likes to treat a substance. When analysis is attempted, the fractions that are obtained are rather impure and complex. Different fractions are called respectively, humin, fulvic acid, alpha humus and beta humus, finally leaving an insoluble humic acid. It seems that an ecologist will have to leave the chemists up their five-branched blind alley, because none of the fractions has been identified as being particularly the *sine qua non* for nourishment of any of the plants or animals. The final fraction often stays strongly bound to clay. This, among other things, allows the Russells to talk of the clay-humus complex, and there is no doubt that the two can be most intimately connected.

We may now turn to consider the structure of clay itself. To begin with, 'clay itself' is an inaccurate phrase because there are several different minerals that give rise to clays, and in addition, in nature, the clays that form the subsoil of fields are often mixtures of the products of different minerals. The properties of those mixtures may well in practice be more important than the properties of pure clays, about which there has been a great deal of molecular research.

Aluminium silicate is the staple chemical of clays, and both aluminium and silicon are elements that can be ions on either side of a salt. They can be either 'cations', as in silicon tetrachloride or 'anions' as in sodium silicate. It is not therefore surprising that clays can serve as a reservoir of other elements as well as of water, giving them up easily, as they may be needed by the organisms of the soil. This is true also for humus. In spite of acting thus as temporary stores, both clay and humus maintain their integrity. They may contain other elements such as magnesium and calcium without altering their essential structure and behaviour very much. Thus, to sum up, it is as if the organisms living in the soil lattice dwelt in rooms in which the walls were built of various foodstuffs which could be taken up, put down or exchanged: not merely a ginger-bread house but a veritable pastrycook's.

In mentioning plants' foodstuffs, we have arrived at what is usually considered to be the beginning of soil science, namely the chemical elements that a plant needs in order to live, grow and reproduce. Three of these are famous; they are known as N.P.K., which stands for nitrogen, phosphorus and potassium. The source of nitrogen can apparently be either the ammonium salt or the nitrate. The plant root appears to cope happily with ammonium and nitrate in very dilute solutions and uses them to build up protein. These three, N.P.K., are well known because they are the main constituents of the commonest artificial fertilizers. All three elements have proved able to foster and increase the crops in the fields to which they have been applied. By growing plants in water cultures and leaving out various constituents one by one, it has been fairly easy to show that plants also need calcium, sodium, magnesium and sulphur, as well as the silicates, and minute quantities of a large number of other elements such as copper, boron, manganese or aluminium, many of which are nevertheless poisonous if present in too-great quantities. These are known as trace elements, and in New Zealand and elsewhere benefit to run-down land has come from providing them.

Since for thousands of years there were no minerals used on soils, it follows that the micro-organisms, especially the bacteria, can liberate the elements into the soil solution for the use of the plant. The root hairs, microscopic and delicate as they are, then obtain from the soil all the nourishment that the plant needs, and they do this because they are bathed in the soil solution which is thinly dispersed on the particles of soil or in the minute pores between them.

Apart from a catalogue of the organisms in soil, what this chapter on autecology amounts to seems to be something like this: when we think of life on the farm, we think first of cows and sheep and grass and grain; but there are others. These are worms and wire-worms and fungi and everything down to bacteria and even phages. When we speak of the salts in the soil solution, on which this chapter finishes, the facts seem to make the smaller of the farm organisms, especially the bacteria, of importance equal to the larger. Their relations, however, constitute synecology, and so fall to be treated in another chapter.

The farm/II/the web of life in the soil 3

Animals and grass
'The cow eats grass,' is a simple sentence but is not a simple statement when it is put into terms of action and reaction.

The cow is an animal bred from ancestors in the wet and lush jungles; she is therefore by nature both a browser and a consumer of long grass and herbs, and it is somewhat unnatural to keep her on short grass. She has no biting teeth in her upper jaw but instead a leathery pad, and she has a broad smooth muzzle which prevents the lower teeth and the pad getting very near the ground. Consequently, she leaves grass as rather long stubble compared with some other animals. The horse for example has cutting teeth in both the upper and lower jaw and so has the ass. They are grazers of dry lands, the horse for the plains and the ass for the mountains. The sheep is also a mountain animal but rather of the lower slopes, compared with its near relative, the goat. The sheep, like the cow, has only a pad in the upper jaw but it has a split upper lip which enables it to get closer to the ground than the cow can, while the rabbit, with both split upper lip and cutting teeth in both jaws gets closer than either. So much for the variety in the grazers; there is also great variety in the grass.

There are about a hundred fairly common species of grass in the British Isles and a score of them are of agricultural importance. A field will contain several kinds of grass, some large, some medium-sized and some small. The effect of an animal such as the cow may be compared with that of a lawnmower when habitually set high, allowing large species of grass to grow, which is undesirable in lawns where usually the aim is to have very small and relatively unproductive kinds of grasses in order to make fine turf. For extreme cases, such as golf greens and bowling greens, the lawnmower is set down so as to cut very close to the ground. If you habitually cut a lawn with a lawnmower set high, you will get a much coarser appearance and much more bulk of grass to carry away than if you habitually cut with the blade set close to the ground. In the same way a field can be more productive of grass when habitually grazed by cows than when habitually grazed by close-cutting animals.

Another valuable characteristic of the cow is that she is less discriminating than several other animals about the taste and quality of her food. This may be noticed when feeding hay in winter, cows will eat and do fairly well on hay that has gone a little musty, that is, growing fungus of some kind. She may prefer and do better on hay that is not musty but she will eat it and live off it, and equally, when grazing a field, she will be less fussy than a horse would be. I have known cows eat a great many buttercups, which are bitter herbs and can be poisonous if there is not enough grass to go with them. I have only once come across anyone who had cows poisoned by buttercups and that was on tip land where there was no grass at all and the cows in their undiscriminating way had tried to get a living off the buttercups.

Where animals habitually drop dung, that patch of the field is described as 'stained', and if you have a large number of the same species of animals there will be an increasing area of stain, which is not properly grazed because the animal will not put its nose too near dung of its own kind. If you mow that grass with a scythe so that it falls a little way away from the dung below it, then the animals will often pick it up and eat it; so, clearly, there is not necessarily a taint in the taste of the grass; it seems rather that the animals will not eat near dung. This applies especially to cows and horses, whose dung stays in the place where it is dropped; it does not apply so much in the case of sheep, especially those on slopes, whose dung is rounder and more mobile, so one does not often see a hillside pasture pimpled with larger grass as one does see lowland pastures that have been grazed by cows only, or with jungles near a hedge left by horses.

A field used only for horses gradually deteriorates, with the area near the dung growing long grass. Long grass in itself is disliked by horses while they can find finer grass elsewhere in the field. They tend to keep such an area for dropping their dung so that the area gets bigger and the sweetest grass is eaten and turned into dung to manure the part that is least used.

There are therefore a good many facets to the statement 'a cow eats grass', and running over them has revealed how very plastic the practical problem really is. The feeding of different stock upon the grass will make a difference to its productivity, and so will the fact that there are so many species of grass, any of which may come in as air-borne seeds and dominate the pasture if the conditions are more suitable for it than for its competitors.

Thus, even before we get down to investigating the soil under the grass, we have a picture of a highly variable situation with possibilities that cannot be weighed or foreseen from detailed knowledge of either the animals or the plants. They have to be studied in that particular field; that is the truth of the matter and must be admitted.

Consumers of roots
Nevertheless, let us continue the general account. On the surface of the field we see new blades of grass constantly being consumed, during the grazing season, so that no blades that were growing in April will be found in September, if all has gone well. The cow eats grass, and new blades of grass take the place of what the cow has eaten. In Chapter 2, we read that the same is true for roots. The replacement of rootlets is invisible and so is not generally realized; in fact it has not been precisely measured for grass roots. Nevertheless, if with a spade you lift a piece of turf and look at the tangled roots beneath you will see relatively few of the little white tips that denote a new and active root, but you may see a great mass of brown rootlets and fibres which are the old obsolete roots. Root formation has been precisely timed for the rootlets of an apple tree, for rootlets which were perhaps a little thicker than carpet thread. Each rootlet developed anew and lived for only about seven weeks. The plant is thus constantly throwing out new rootlets to take the nourishment from pieces of soil that are not already occupied by rootlets, and there is quite a mass of old rootlets to match the shortlived blades of grass that are consumed by animals on the surface of the ground. One of the important qualities of roots is evidently their continued activity.

We know from Chapter 2 that the principal consumers of these roots are the wireworms, which are usually considered to be pests in the soil because if you suddenly cut off the supply of grass roots by ploughing the land, and put in potatoes, then undoubtedly these will suffer very greatly from wireworms, who have been robbed of their normal food. Remembering the example of the apple rootlets, most of the grass roots eaten by the wireworm will be freshly dead and due for consuming, just as the grass blades are consumed on the surface above, so the wireworms clear the soil for new and fresh rootlets. This is obviously very important work, but the wireworms are not alone in this. There are hundreds of species of important organisms under the soil and the exposition of their significance could be cumbersome. It will perhaps be well to consider further the work of wireworms and shall we say earthworms as examples, leaving the smaller multitude of Chapter 2 until we see how much detail of their activities seems really necessary to the understanding

8 Cultivators. Soil is always on the move and its movement will bring down a stone wall in a century. All soil organisms burrow: earthworms with moles seeking them; centipedes; beetles and their grubs; leatherjackets and mites. Some mites carry soil about on their backs. Roots and swollen crowns of herbs also move the soil: burnet, clover, and chicory are shown in the drawing. There are also the fungal orders, and the surface-breakers, such as starlings, crows and rooks.

that we are seeking.

We are beginning with wireworms, with which go many other kinds of insect larvae. The wireworms as we saw are the larvae of the click beetle. They exist in enormous numbers. One might expect that the world would be overrun by adult click beetles, but this does not happen, probably because some birds, especially crows, starlings and so on, work over the fields and fill their crops with wireworms. As they do so they deposit dung and so some at least of the nourishment from the old grass roots goes back again on to the surface of the soil and washes in. Also, when the birds work over the fields in this way the surface of the ground is being given a treatment comparable to that of the spiky aerating roller that is used on lawns, so that the rain finds little holes or their traces made by starlings' beaks into which the water can sink and so seep down to the roots.

The relation of wireworms and starlings came to me when I happened to see the beak marks about two inches apart on a small mound of soil after the starlings had passed. It was, I think, a mole-hill that had had a wheel over it. The most famous pastures in England, the classically successful fattening pastures of Romney Marsh, give one of the highest wireworm counts of any in the United Kingdom.

It is to be feared that much manurial value is lost when starlings roost on city buildings and other places where their droppings do no good, but this is perhaps not a large proportion of their droppings during twenty-four hours. Domestic hens drop a good deal of dung on the pastures they are picking over and we may presume that starlings do the same.

Thus we see the wireworm-starling combination doing two important services: cultivating the surface of the field, thus improving its physical condition, and also returning some of the nutriment, derived from obsolete roots, to the surface of the soil, where it can serve as valuable manure. I praised the wireworms in a book *Soil and Sense* (1941) but the point was not taken up and by 1969 there was more to say.

Mulch: earthworms

The beneficial activities of earthworms overlap to some extent with those of wireworms and starlings first described.

Wireworms and starlings are concerned with obsolete roots underground, but on the surface of the soil lies a certain amount of obsolete grass, and we should consider what happens to that and to the tree leaves fallen on the surface. There are also the dung droppings. It is well known that the earthworms help to take these wastes down into the soil, dragging them down into their burrow. If the cow pats are lifted, the openings of worm shafts can often be seen.

So our understanding of the role of wireworms and earthworms is thus, that whatever they may supply in the nourishment of plants, they have also a great deal to do with maintaining the open texture of the soil. In roughly the same way, many, perhaps all, of the organisms in the soil have this double role; they break down vegetable remains for the nourishment of new vegetation, and, because they are alive and striving, astir with activity, they keep the soil moving and open: they cultivate it (Fig. 8). This has been touched on in Chapter 2, but it can hardly be overemphasized.

The cultivation by organism is not a mere improvement, but a major operation. Because of their multitude their cultivations have great magnitude, which is shown in a pointed way by a modern cure for apple scab, involving earthworms as principal agents. For many years owners of orchards had great difficulty in keeping down a fungus called the scab which makes patches on the leaves and marks the fruit very badly. Owners were in the habit of spraying the trees for scab repeatedly, up to eight times during the season and then not mastering the scab. However, scientists at last solved this problem in a rather unexpected way, which is entirely in keeping with our ecological view of husbandry. Knowing that when the apple leaves fall they have on them the scabs, which will shortly liberate spores to blow about and reinfect the apple trees, men have found that if they spray the fallen leaves with something that attracts earthworms—namely urea—the worms will attack the leaves with avidity and drag them down into their burrows before the spores are liberated during the winter. Practical men to whom I have mentioned this are inclined to say, 'At last: the scientists are seeing that they have to fit in with nature.'

So it is in mass action, no less, that worms do drag down vegetable remains from the surface, making hundreds of vertical taps to let rain and air down into the soil. When the worms get their food down, they mix it with soil and a little limey material which they secrete from special glands, provided that there is sufficient lime in the soil. Worms pass this 'mull' as the mixture is called, through their bodies and about a half of the worm species leave casts on the surface of the soil, giving a surface dressing of humus.

Acidity and oxidation balance, pH redox
Earthworms can secrete lime to go with their mull only if there is sufficient lime content in the soil, and in fact you do not find any quantity of earthworms in a soil of high acidity. This brings us back to the question of the species of grass because the productive agricultural species tend to be found mainly in the range of low acidity, that is fairly high pH number. Timothy or catstail is one that can tolerate more acid conditions than most other agricultural species. The grasses found in very acid soils are the bents and the wavy hair grass familiar on mountain swards. They are almost unproductive and will keep say only one sheep per several acres of springy pasture instead of several sheep on an acre of firm and open-textured sward.

In a productive field the species of grass or herbs sown may be cocksfoot, rye grass, timothy, red and white clover, lucerne or sainfoin, and good kinds of meadow grass and other species; but merely sowing those seeds, however good the strain, perhaps even bred at Aberystwyth, will not ensure their success. The field in a year or two can easily become all bents again brought in by the wind, if for the better grass species the conditions are not right. One of the more important conditions is the hydrogen-ion, as it is called, concentration, which expresses the acidity at that moment. Springy turf underfoot is a sign of acidity and is due to a velvety mat of grass decomposing very slowly on the way towards peat. In acid and in airless conditions, the mat is not freely attacked by earthworms nor even by bacteria. To express hydrogen-ion concentration, scientists have a notation that was not designed for easy understanding by laymen. Doubtless the notation could be improved, but a change would be very trying for all practical men who have become accustomed to the present notation. When the practical man sees the shorthand sign pH he knows that if the number that follows it is less than 7 the soil is on the acid or sour side of neutral. This does not matter for most vegetation provided that the figure is not below 5. If it falls to 3·5 it is at the limit for even quite acid-tolerant, plants, for example sallows. If, on the other hand, the figure is above 7, the soil is on the alkaline side and if it exceeds 8 plants are getting into danger again. It would not be right to sacrifice this familiar understanding by trying to make the notation less scholastical.

The use of the word acid needs a little explanation. Before pH was understood chemists had measured acidity or alkalinity by analysis with strong chemicals, but that would be quite unsuitable for pH, which has a much more delicate and ephemeral quality. It is measured by electrical conductivity or by the use of dyes which show different colours at different levels of pH. By either method the chemical action of tests is so little that it can be taken as negligible. It can be compared with testing the sourness of an apple by licking it. The slight alkalinity of the saliva would indeed change the pH at the surface where the contact was made, but this change would be so minute that we would not have any doubt about saying that one apple was more sour than another.

In the 1920s the work on pH rightly made a great impact on the thinking ecologists and I think that perhaps a version of the Russells' Table 119, which summarized the work of O. Arrhenius on the soils of two hundred Swedish farms, is worth reproducing here (Fig. 9).

Doubtless there are many ways in which the acidity of the soil solution can operate. One is that too low pH number makes aluminium and manganese more soluble. These are harmful substances to which plants will be susceptible in various degrees. The pH is constantly undergoing changes due to various activities: such as the liberation of acids in the breakdown of humus and the absorption or liberation of the weak acid formed by carbon dioxide, which is the principal product of respiration. However, to make a fundamental change in pH it is usual to dress the soil with lime which, being a mild alkali, can correct over-acidity.

We can place hydrogen-ion concentration with another important condition of the soil called the 'redox' potential. There are numerous reactions going on in the soil that are in the nature of a balance between oxidation (more oxygen) and reduction (less oxygen relative to hydrogen). These may safely be said to be the work of micro-organisms. The Russells mention the system nitrate-nitrite, sulphate-sulphite, manganese dioxide—manganic ions and ferric-ferrous ions among others. The condition of oxidation—

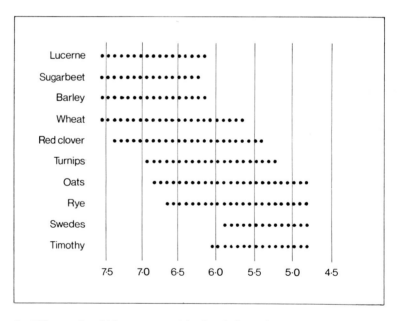

9 PH range in which crops occur. The dots indicate the pH range in which the crops occurred, even if they gave poor yields, and it is possible to see lucerne belonged to low acidity, that is a high figure in the pH, and that timothy was a favourite at higher acidity than most, i.e. at a low numerical figure of pH. I have slightly altered the diagram given by Russell because I am anxious that it should convey the area of tolerance of pH for the crop in question rather than be read as showing the best pH at which to grow it.

reduction balance may well prove to be more fundamental than pH, controlling as redox does the process of respiration, which provides the energy of living organisms. But historically pH was discovered first and Pearsall's addition of the importance of redox entered the field partly as a criticism of too great emphasis, he thought, by E. J. Salisbury on pH alone. Pearsall in the *Journal of Ecology* in 1926 recorded his observations on flourishing vegetation at low pH numbers. For example, *Glyceria aquatica*, which blocks the disused Bolton canal in Salford, is a sweet grass that grows in sour conditions.

Pearsall loved walking on the Yorkshire moors and fishing in the Lake District and had marvelled at the success and vigour of bog and lake side vegetation. Over the years he worked to correct the impression made by pH alone by making a collection of about 300 samples, and electrically testing for pH and redox. That both are intimately connected with living organisms is shown by the necessity of preserving samples in toluene if the values are not to change with the hours of keeping. The general result of Pearsall's work was to establish that there was a neutral zone of oxidation and reduction lying between 320 and 350 millivolts, just as the neutral point of pH lies near 7.

Among the various reactions I have cited above from Russell's book, it was the ferrous-ferric that Pearsall took as indicator of a soil's redox condition. This was a sound choice; partly because it acts as an indicator, making the soil blue black in the reducing state and rust-red in the oxidizing, and partly because it is so rarely absent, and partly because iron (ferrous) bacteria form pans below the active layer of soil.

Pearsall's other tests in parallel were of bases, replaceable iron, free lime and nitrate. The work yielded in the first instance three papers, published in the *Journal of Ecology*, Vol. 26, 1938. They formed Professor Pearsall's presidential address for the year 1937. This first paper is on the theory of oxidation—reduction potentials, the second on woodland soils, and finally one on moorlands and bogs.

In the woodland soils the typical birch wood had pH less than 3·8, very definitely on the acid side; no nitrates were present in the soil, nor were there any earthworms. But above 3·8 pH he had both nitrates and earthworms and up to 5·1 pH this was characteristic of oak woods. Above 5·8 pH he still had nitrates and earthworms but the woodland characteristic tree was ash. Turning to the moorlands, we have soils that gave him less than 320 mv redox potential: he called them reducing soils and found no nitrates there. Over 350 mv. he called them oxidizing soils, which had nitrates, forming part of the aerobic pattern. On draining, these moorland and bog soils became oxidizing and they also developed an acidity, as the 90 per cent organic matter that they contained became oxidized. Pearsall gives an interesting comparison about soil in which the grey sallow, *Salix cinerea,* was growing. The driest of such soils had redox of 325, whereas the wetter ones were 163 and 48. Here we encounter that interplay of draining and base content and fertility which is known in empirical and practical studies, which I have tried to illustrate in Fig. 10. The interplay is further shown by his observations on *Molinia,* which is the moor grass that forms an important constituent in the diet of mountain sheep, according to the work of Sir George Stapledon described in his book *Hill Lands of Britain. Molinia* can appear as a prominent member of a plant community under two conditions, either on reducing soil, provided that they are above 4·4 pH (approx); or oxidizing soils even if the pH is below 3·9. Pearsall concludes that acidity may be supportable provided sufficient oxygen is present.

Discussion of Pearsall's work gives an insight of pH and redox as current-account assessments of soils, leaving open the question of longer-term causes. It is for instance probable that part of the acidity of moorland soils is due to the method by which they have been farmed. Year after year lambs leave the hill country containing phosphate and lime in their bones and little or no phosphate or lime is put on the hills in its place. Such at any rate was the theory of men like Stapledon. Dressings of basic slag, which contains both lime and phosphate, can make an enormous improvement on upland pastures. Thus, theory seems to be half borne out by practice, but there is little hope of separating in practice the effects of shortage in each component, lime and phosphorus. With them also are involved the 'trace elements', which leads to consideration of minerals generally.

Mineral efficiency: nitrogen
In the course of nature the soil obtains its supply of minerals slowly from the subsoil, by the agency of trees, bushes or deep rooting herbs whose roots go down to the subsoil itself. By the acid in the roots' respiration or that of mycorrhiza, the minerals are dissolved in the soil solution. Here then is another major service performed by roots, through their respiration, as if it were a by-product of their search for energy. For this process, as for so many, it is important that the soil texture be open and aerated. The roots must

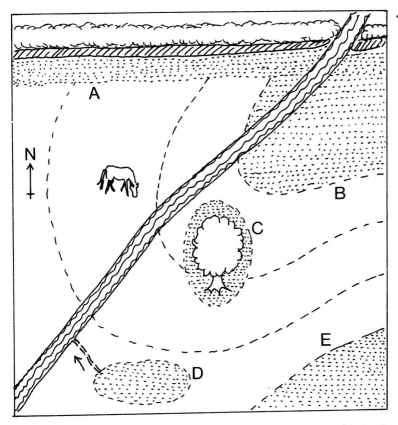

10 A faulty area of grazing. Two ponies on my pasture near Lowestoft in 1936 showed up unpalatable parts, all of which an old stockman called 'sour' but for various reasons. I think they were as follows, giving the explanations and (in brackets) the test called for: (A) tainted by rotting humus in the choked ditch (redox and pH); (B) wet through ditch choking (redox); (C) sour because shaded (redox and pH); (D) and hollow improved by draining (redox); (E) lime leached away (pH).

be able to go down, if they cannot do so the pasture may deteriorate to peat.

Assuming, however, that all is well, we can find useful forage herbs like dandelion or ribwort plantain, which animals like to eat and which have roots going down very much deeper than the grass roots do. It is they particularly which, as they breathe excrete the carbonic acid, which acts on the rocky materials of the soil and so liberates the minerals that are required for plant growth. Plantain, dandelion, chicory and burnet are not so well known as the leguminous herbs, such as lucerne and the clovers, which in addition to their mineral efficiency—the technical term for the process just described—house bacteria with the capacity of liberating nitrogen into the soil slowly while they are alive and more swiftly as the roots become obsolete.

There are two well known leguminous plants which are rather interesting in that they must have an acid soil in which to flourish; they are the broom and lupin. Most other nitrogen fixing plants like to be at the high voltage of the redox scale and the high number end of the pH scale and the same is true for the bacteria of the *Azotobacter* group, which are free-living.

In Chapter 2 we saw that there are fungi and blue-green algae that can perform the same service. Furthermore, lightning causes nitrogen to be oxidized in the air and carried down in rain, but this does not amount to more than a pound or two per acre per year. Nitrates leach out of the soil very easily and it has become the practice for farmers to replace them regularly, for example in top-dressing grassland say in March, which enables them to get their hay crops earlier than they would if they waited for the grasses to grow while taking nitrogen up slowly from the soil. My neighbours in the north-west, where hay weather is very uncertain, have found over the years that it pays them to give the land this top-dressing.

When nitrates are thus provided for the soil, the supply does inhibit the action of the natural fixers of atmospheric nitrogen, but otherwise it does not seem to upset the economy below the turf and doubtless the nitrogen fixing agents can resume their normal duties when the added nitrate has gone, which happens fairly quickly. The main advantage of the natural agency is that it will come in to correct any shortage, it being the nature of microorganisms to do what is required without orders from human managers.

Soil organisms in action
Study of action of wireworms and earthworms has already suggested that fertility depends upon the soil being sweet and aerated. We are now to carry the questions of actions harmful and helpful down to the smaller organisms detailed in Chapter 2.

The springtails, as far as we know, do not attack the crop but help to keep the soil on the move. As to mites, not forgetting their special services in soil cultivation, it must be admitted that some species may attack crops. The same is true of nematode worms, although doubtless they also do a great deal of good in clearing up scrap vegetable matter. The action of protozoa was once thought to be mainly harmful because they consume bacteria; however, it has since been found that many bacteria flourish in the presence of protozoa, although the protozoa are consuming them all the time. Fungi and bacteria have a wide range of activities, from useful breaking down of decayed vegetation to more or less severe attacks on plants. It is not possible to define too closely the harmful or helpful or neutral action of a particular breed of bacteria or of a particular strain and to some extent the same is true of fungi.

This subject is very new and in a considerable state of flux, but there seems no doubt that not only are the bacteria and fungi extremely versatile, as related in Chapter 2, but that variation can take place within a few generations and fungi can even affect each others' physiology without, as far as we know, breeding in any way comparable to that of higher organisms. Smaller and even less well known are the phages and enzymes on which the health of bacteria depends; these will provide a fruitful field of research for many years to come. Their structure is beginning to be revealed by the electron microscope.

A generation ago there was a tendency to jump to conclusions such as the one mentioned that protozoa were entirely harmful due to their consumption of bacteria. It was by no means clear then that bacteria, fungi, mites and nematodes could be both helpful and harmful in the web of natural fertility; but research has given a clear picture of great plasticity and which should induce a healthy reluctance to draw conclusions from technical knowledge alone. Perhaps, however, by proceeding cautiously we can attain a general grasp of the usual situation. First, we may remember from Chapter 2 the simple algae growing on walls and rocks, protected from drying up only by a thick cell wall, and the lichens or mosses with better water storage in their own bodies, and comparing them with the invading grass seedlings sending roots down into the moss and dust sponge. At that stage they possess soil, of a poor kind, but soil to conserve and feed them with moisture. In fact soils may hold 5 per cent of their dry weight in water in the case of sandy soils and up to 60 per cent in clayey soils. Thus we have come back to a fresh statement of the essential quality of soil, as a water store on which higher plants rely.

Evidently clay soil is better water holding soil than sandy soil, about twelve times as good, but clay itself would not make good soil because when it dries it forms a hard material which would be almost impermeable to the next shower of rain and would not accept roots. Roots cannot permeate solid clay; they must have cracks and spaces. Consequently, one of the most valuable features of a soil is that it should dry to a crumb structure. The clay itself has the property of being flocculated, that is drying with spaces in it, under the action of certain chemical substances, but the processes controlling flocculation are not fully understood. The part played by vegetable remains, commonly referred to simply as 'humus', is better known and it can be quite dogmatically stated that humus makes heavy soil lighter, and that is, easier to lift on the spade and more likely to have a crumb structure. Humus also makes light soils heavier. This is because the rotting vegetable remains are to a certain extent slimy or spongy, or both, so holding water, taking the place of clay up to a point in that respect. The pores and cracks in a crumbly soil allow the rain to go down and to be stored for the ease of roots as they want them. If the soil is to retain its crumbly structure in spite of the rain, the crumbs must not turn into mud on being wetted; they must be to some extent what is called water-stable. In fact, these crumbs, which vary from the size of a grain of wheat to that of nuts of various kinds, are usually fairly stable to wetting, owing to the fact that they are bound together by some of the soil micro-organisms, particularly some of the soil fungi. These micro-organisms live on vegetable remains such as the remains of roots.

Reverting to the proviso that roots can penetrate the soil only where there are cracks for them to enter, we should also remember that in addition to requiring water in the soil they require air, because like all living organisms they must breathe, that is, burn up some of their tissues, for the sake of obtaining energy, using oxygen of the air for the purpose and making carbon dioxide and water. So we can visualize the soil as a lattice of particles in which roots can freely wander and breathe by means of the air and can suck up water for the use of the plant. The water therefore must not abolish the air and rainwater does not usually do so. Raindrops carry down bubbles of air with them and have plenty of air dissolved in them as well, whereas water seeping up from below usually abolishes the air, making the lattice waterlogged, whereupon the roots of most kinds of plants will die very quickly for lack of air.

Drainage is thus of great importance in the air supply and a large part is played by the roots themselves in cultivating the soil. From one half per cent to three per cent of the soil under turf may be formed by decaying organic matter which has very largely come from the roots of the turf itself. It may not be generally realized that much of the decayed and dying vegetation in the soil is freshly put there.

Root action can be astounding in magnitude. In sandy loam in Saskatoon, spring wheat plants sown in the ordinary way in drills 16 inches apart and spaced 18 to 20 plants to the foot had half a mile of roots for each plant. An isolated wheat plant on the same soil had produced forty to fifty miles of roots in the eighty days of its life.

In general a rough estimate would be that one-eighth of the crop is left as residue in the soil, namely roots that have finished. To these are added plant remains from above. We saw how active worms can be in taking down dead grass, fallen leaves and dung, as are the well-known burying beetles which take down carrion.

Considering all that we have managed to include in this section, and there could be much more in support, there seems to be one special conclusion: that perhaps the major importance of all those organisms that live on dead and decaying vegetable matter, or are carnivorous on others that do so, lies in their constant movement and aeration of the soil, thus maintaining the spongy or lattice structure—like a wet sponge or a dried sponge, but a sponge. We concluded in Chapter 2 that the sponge was equal to gingerbread or better, so far as its usefulness went for plants via the other organisms.

Cycles

We have now gone over the second level of ecology in the farm, namely the synecology, and we may say one or two things with certainty. One consists of some mathematical common sense arising from the fact that we are undoubtedly dealing with a main cycle of events in fertility, decay and regeneration.

Although many details of farm ecology are recently learnt and strange, the system, the synecology, has been known in outline for several generations: the cycle or the 'wheel of life'. Cycles are characteristic of other eco-systems too; but the farm, being near home, will perhaps serve best as a model with which others can be compared, whether lakes or forests or seas. Let us construct the cycle from what we have learned so far, with the help of Fig. 11.

In the spring the grass grows up and the roots grow down. eventually to decay, along with the dung that the animals have dropped. Thus all plant remains are continually being converted into smaller and smaller pieces. Earthworms play their part in continually dragging dead vegetation into their taps and burrows. Eventually, when all the agents have lived on the 'humus' there are the mineral salts to be absorbed by the root hairs of the grass and so help to make new grass for the animal to eat.

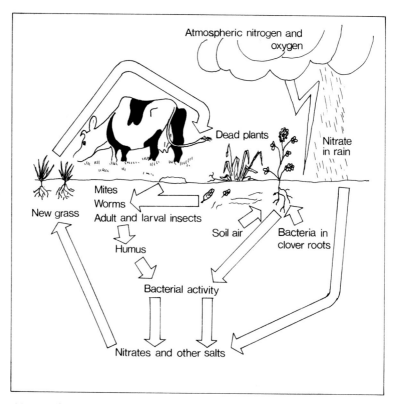

11 Fertility cycle. The fertility cycle is based on humus—the store of wastes and food. The cycle of nitrogen is shown, but all the other fertility materials also circulate through humus, aided by clay. The supply of nutrient in rainfall is usually small. After Dr. Stanley Frost.

A cycle can go faster or slower. There is a cycle even without the animal, but a slow one, when dead blades of grass fall and lie on the ground, forming a 'mat'. Even this mat is eventually destroyed by insects and bacteria, but the process is a slow one, because the weather may be too wet or too dry and because a mat is not well aerated. However, in time the brown rotting substance, which is like peat, becomes humus and part of the humus becomes salts to feed the new grass growing up through the mat. This new grass is of those poor kinds which can stand the airless, acid conditions: thus, in these conditions, there is little grassy substance made each year.

The cycle goes faster if grazing animals are let on the grass. The grass is now half rotted by the time that the remains leave the animals' bodies. The dung itself contains salts and nearly all of it is fit for the insects and bacteria to turn into humus and earthy substance within a few months. Nourishment from urine can act even more quickly. Within limits, more cattle on the land means an increase in everything else, more grass and more milk and a faster cycle per annum. Up to a point, all the processes can be made to go faster. The more the cattle the more the dung. The more the growth of roots to nourish the meadows, the more the carbon dioxide from their respiration to liberate salts from rock particles and so the more the new plant growth, to feed more cattle: so we come round in the cycle. The limits are technical, such as the more the cattle the more the density of their parasites on the ground, and that, however much the carbon dioxide, the decay of rock particles is rather slow. However until those limits are reached, speeding up the rotation of the cycle is merely a matter of applying more human energy to caring for more cattle in the winter. Moreover, in practice, there are biological 'savings' because it has been found that humus and combined nitrogen accumulate under grassland.

We may call the cycle outlined the 'fertility cycle' and we may note that it can be measured in cwts of beef grown per acre per annum or in money per acre per annum: which shows that fertility is a rate—not a mere quantity.

It is evident also that the 'humus' in the fertility cycle is working capital and that crop is the interest, or it might be put that the humus is the 'variable' and the crop the 'differential coefficient'; but the working capital and interest is probably a more acceptable version. Whatever model is chosen, however, let this be clearly appreciated: fertility is not magic on the one hand, nor material on the other, it is live maths. It is the kind of common sense 'horse' mathematics, which alas the public is rather loth to apprehend— except where money is the expression of it, when it often seems to 'come natural' to people to understand it.

Arable husbandry

The great and widespread way of speeding up the fertility cycle is by arable farming: using the plough and all that has gone with it over the centuries. When the soil is disturbed or turned over to ploughing depth say four to eight inches—air is let down to deeper layers, which means an increase in the activity of many of the micro-organisms of decay. Decay of humus is speeded up. Consequently heavier crops of cereal can be grown. Cash can come from sales of corn, and still, as part of the system, more beef can be raised than on the same acreage of grass. The beef is fed on root crops and on clover hay, grown as part of the system: all that is understandable by horse mathematics, the speeding up of decay processes by human husbandry, in systems that have been developed with increasing skill down the centuries. That history continued into the nineteenth century and well into the twentieth: the classical example is the Norfolk four-course shift.

Whereas grass farming, even when done lazily, shows gradual increase of humus in the soil, arable farming tends to reduce humus by the cultivations, unless it is restored in carted on manure and in growing grass and clover for hay and in grazing the aftermath. These provisions made the systems viable over so many decades and were legally enforced in England until 1926.

Thus we have been discussing farming up to about 1926, although it has been illuminated by research since. Of developments since 1926, three are to be considered especially.

The first is the spring dressing of grass with nitrogen manures, mentioned already, which is practised by my neighbours in south Lancashire, and I believe nearly everywhere else, and seems to be beneficial in getting a larger hay crop sooner.

The second matter is called lea farming in the north of England, ley in the south and pronounced accordingly. We have noted that grasslands tend to accumulate humus and that arable farming, unless it is carefully watched, tends to lose humus, and that humus is, properly speaking, the working capital in the soil. There is no point in accumulating it unless it is giving a better crop every year. That would be storing away capital that should be working. Losing humus, equally, is like losing the working capital. What has just been written of the humus content applies also to the content of nitrate in the soil but it has been mentioned that nitrate is the nutrient salt that washes out most easily in the rain, which is called leaching.

For a long time in some parts of the country, near the Scottish border for example, farmers have made the best of the humus changes by using the system we are considering, often called alternative husbandry. A field is left in grass for four, six or more years to gain the benefit of animal grazing, alternating with a period under the plough to cash the benefit. The working capital in the soil is thereby kept at a reasonable level on the average. Due to periodical ploughing, the pasture does not become stained, and there are not as many parasites as there often are in permanent pastures.

In the early part of the century, R. H. Elliot of Roxburghshire, advocated alternate husbandry as a general practice, and the plea was taken up later by Stapledon, who succeeded in having it made official policy when the Second World War was imminent. It meant that many unproductive grasslands were made 'productive' by ploughing them up. Unfortunately there was not then or later any incentive to lay the arable land back to grass.

When I visited Clifton Park at about the time of the second war, I was disappointed to find that the fields on which Elliot had practised the system with so much enthusiasm were not being used in that way by his successors. It is true, however, that the method is not a universal panacea. Apart from the extra work involved, there are fields where permanent pasture is more suitable, for example where land is rather wet and poorly drained, making it difficult to work as an arable field, and where temporary grass may not give such good results as a permanent pasture in which the species and strains of grass have developed to fit the circumstances, and to give sweet grazing in spite of the poorly drained soil.

That brings us back to Pearsall's work on 'redox' which yielded the third of the developments to be noted in this chapter, namely the 'Japanese' improvement in the cultivation of rice. This has two claims to ecologists' attention: it enhances the crop of the staple food of more than half the world's inhabitants, and the poorer half at that; and it reveals the subtle action of micro-organisms in relation to redox, which is doubtless merely an extreme and well-authenticated example of a whole family of reactions in the soil for other crops and for wild vegetations, both terrestrial and aquatic.

We saw how Pearsall roamed the moors and the Lake District, so getting a keen appreciation of the rapid and useful growth that plants can make in waterlogged soil in spite of the anaerobic conditions. When he followed this work up he saw that it had an application to the growing of rice, which is planted in soil that is subsequently flooded. There are other ways of growing rice, but that is the example we shall take. The rice soil, called paddy soil, is dark coloured below and rust coloured on top, showing that it is anaerobic, from the presence of ferrous iron, below the surface, and aerobic, containing ferric ion, near the surface. It seems that if manure is spread on the surface, it is of no avail, and in certain areas it has been traditional for a long time to plough all the manure in below the surface of the soil. This may seem peculiar, because the anaerobic conditions would not appear to be favourable for plant growth, but the explanation seems to be that if nitrogenous manure lies on the surface it is oxidized to nitrate by the appropriate bacteria and the nitrate immediately seeps into the anaerobic layer where it is reduced to nitrite and finally to nitrogen itself, again by the agency of micro-organisms. By this process, the nitrogen is lost from the system, in fact up to 70 per cent of it can be lost in that way. Following on from Pearsall's work and Mortimer's, Japanese authorities therefore recommended that the manure—it may be sulphate of ammonia—always be put in the deeper layer. Then nothing much happens to it by the action of bacteria until some oxygen gets down into that level too. The oxygen gets down through the air spaces in the tissues of the growing rice plant, thus reaching the rootlets and to the root hairs. There the soil on the roots is often rust coloured owing to the aerobic effect carried down. Then the manure that has been preserved intact is oxidized but not lost. It is absorbed immediately, being as it is close against the root hairs. Thus comes yet another vital action of roots of a special class, namely oxidation.

The effect of Pearsall's interest was that Japanese workers proved through extensive trials that manure should be put into the deeper layers, and the land to be flooded within two or three days if possible so as to preserve it until the roots could take it as they want it. Thus was an old and localized practice given greater sanction and more widespread use.

In this chapter we have referred to three new, or renewed, agricultural practices that are understandable in the light of the synecology. There is the spring dressing of grass with nitrate after the losses in winter rains; then the expanded practice of alternate husbandry; and the 'Japanese' method of growing rice. It may be wondered why no credit is given here to artificial 'fertilizers' in general, or in other particular applications. That raises a large question, falling under the heading of macro-ecology, which will make the third of these chapters on the farm.

The farm/III/the great debate   4

Of chemical manures
'The first time I used artificial manures, I left half the field untreated and I could see no difference afterwards.'

'Yes,' said a neighbour, 'so did I, but over the years I have found it pays me to dress the grass with nitro-chalk just before the spring growth.'

Synecology provides the explanation that those particular failures were due to acidity in both cases, but my neighbour did not give that explanation. To him, and to many others, artificial manures form a belief; in their view a good farmer is one who uses artificials 'to maintain fertility'. The same view is to be found in gardening. An article on gardening may be headed 'priority for fertilizers', and paragraph by paragraph, the writer guides the reader towards many of the products advertised on the same page as the article. This is of some importance, because gardening is as near to husbandry as a great many members of the governing classes reach.

The use of nitrate by British farmers was still increasing in 1968.

Indeed, in Chapter 3, it was concluded that spring dressing with nitro-chalk was sensible and useful; so are leys (south-country spelling in the present chapter) which are generally established with a dressing of artificials; and to grow rice with ammonium sulphate below the oxidizing layer was an able and legitimate method.

Nevertheless, the matter is debatable and is debated.

'Ah,' says the allotment holder, 'you can't grow anything without it.'

He is not referring then to artificials, but to the load of muck that my horse is pulling up the hill. In this modern age, in the year 1969, I sold about 30 loads of muck, and had to end by refusing many more, to poor men—not those who could afford to spend imprudently. A second farmer neighbour, looking at my heaps spaced out over the field, asked, 'Are you on this organic farming?'

'I am inclined that way, but I'm not bigoted about it. I never have been; thirty years ago I wrote a book about these farming questions.'

'You should write another.'

History of controversy

The debate about artificial manures is well over a hundred years old. Authorities at the beginning of the last century had observed over the years that the more humus there was in the soil the better the crops were, and they therefore believed that plants fed directly upon organic remains. As a matter of fact, plants can do so, to some extent at least. In 1946 the *American Journal of Orthodontics,* a publication that is rather remote from most plant physiologists, published a report by Francis M. Pottenger. He grew beans in sand saturated with excreta of cats from a feeding experiment. The sand smelt strongly of the organic waste product called indole, and so, when they were harvested, did the dwarf beans. This is referred to in more detail by W. A. Albrecht (1956). It deserves to be better known and followed up. Albrecht also refers to absorption of carbohydrates, mentioned in Miller's book *Plant Physiology*.

It was a considerable feat for the scientists to convince the world that the food of plants consisted of the such simple substances as the carbon dioxide of the air, water and salts such as nitrates, phosphate and potassium.

The hundred-year-old debate has been settled thus: in 1840 Professor von Liebig emphasized farming by chemicals; in 1970 Henry Fell and other farmer ecologists said that that was quite correct but dangerous in practice. Here the story is put in a nutshell.

Liebig derided the humus men quite out of fashion, notably in his reports to the British Association for the Advancement of Science in 1840. This greatly influenced the advancement of scientists and heralded a full century of agricultural science as we have known it, namely chemistry.

The claims of the chemists were furthered by plot experiments on fields. In a few plots such as the starved ones on the Broadbalk field at Rothamsted, the crop managed without any organic manure being added by man, and thus appeared to favour the mineral theory. Those were the trials that received world-wide publicity whereas on many other plots growth without dung broke down after a shorter or longer term of years. In recent decades, however, it has been realized that oak leaves were blowing on to the Broadbalk plots, bringing humus etc.

In the short term, there is no doubt that minerals as a supplement to humus often increase the yield, and the organic protagonists have probably failed to give chemicals enough credit. The credit, however, is limited by the phrase, 'supplement to humus'. The outstanding question for many years was whether chemicals could form a substitute, for which question a short run of years is no test, because of humus residues from earlier treatments, and of the one-eighth or so of the crop left as residues in the soil.

There have always been men who thought that purely chemical farming was folly, notably R. H. Elliot, the Scottish landowner mentioned in Chapter 3, who also farmed and called attention in an attractive and fascinating way to the value of deep-rooting herbs. He thought that ley farming was better than chemical farming by a long chalk, but he did in fact use a few chemicals to start his leys, which shows that in practice he was not bigoted either.

There were also men within the company of soil scientists who

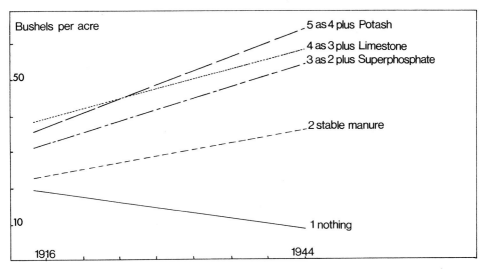

12 Soil deficiencies. These data are rare of their kind. They are from Kentucky Experimental Farm, and are given by Albrecht, (1956). In soil that was not fed at all, 'ear corn' yields fell to nearly nothing. When fed, the yield built up over the years. All but the 'nothing' plot had 'stable manure', but addition of each *chemical element* made an improvement. 'Stable manure' may be misleading. In manure-*heaps*, N.P.K., etc. liberated in rotting are taken up in loose combination with humus, and most are safely held until demanded by the bacteria root-hair complex of the crop. Humus has been found to act as buffer for P.K., etc., but the combinations of N. are too many for the action to be so defined. Nevertheless, it is probably similar. I suggest that, unless the muck was properly heaped and rotted, the nutrients, N.P.K., etc., could be lost in rain as soon as they are liberated by bacterial action. If so, these tests need repeating.

were not satisfied with the disproportionate emphasis on the inorganic chemistry. Sir Daniel Hall repeatedly stated that the physical state of the soil was more important than the chemical, and Dr. E. John Russell in his preface to the great Rothamsted book mentioned already, *Soil Conditions and Plant Growth,* wrote that there had been a great deal of misquotation out of context, of what scientists had said, and statements that had been simplified for lecture purposes had got into the literature without the qualifications that should have gone with them.

It was, I suppose, in the 1930s that anxiety began to reach the public. People came to know of the American dustbowl through Steinbeck's novel *The Grapes of Wrath* and it did cause them to wonder how much of the failure there was due to the attempt to take crop after crop without any of the precautions concerning humus content which European farming had adopted over the centuries.

Precautions about the humus content of the soil were part of the peasant way of life, varying doubtless from place to place, but as a general rule including rotation of crops and restoration of residues and dung. Not to do so would be accounted sinful and robbery of the children's heritage. The attitude and way of life has been well pictured in Philip Oyler's book, *The Generous Earth* and its sequel. Philip Oyler was one of a group of men who called themselves the Kinship of Husbandry and were brought together by Rolph Gardiner, a landowner and forester who was their energetic convener. To me it seemed that the dozen of us in the Kinship were shouldering the lost responsibility for the care of England's soil, which had passed out of the care of the landowners in 1926. Up until then it was a vested interest for the landowner that farmers should maintain the fertility of his soil by rules he included in the leases. Their observance affected the rents that he could obtain by letting the farms subsequently. Mr. Lloyd George, as part of his attack on the power of members of the House of Lords, succeeded in getting Parliament to pass an Act ending the enforcement of the traditional provisions for the health of the land.

The muck and magic school

The Kinship group published *The Natural Order*, a book of essays, in 1945, and on reading it again recently I find much that is truer than ever. It is concerned more with people than with land and it is not by any means a scientific treatise. I doubt if knowledge was clear enough at that time to have let us agree on any scientific thesis.

I remember one incident that illustrates our dilemma in the Kinship. I had guided Philip Oyler to see Adrian Bell, who was farming a near derelict farm in Suffolk. He had taken it over out of patriotism during the war at the instance of the War Agricultural Committee. Close where he was working, he saw a patch of the field about a yard across where the beet were twice as high as in the rest of the field. We stood and looked at it, we three members of the Kinship.

'That is where the fertilizer bag stood,' said Adrian, 'You see that the War Ag. would say that if I'd used much more artificials there could have been a much better crop.'

Philip said nothing but tried to stir the soil with the point of his shoe. The surface did not break.

'Well, I think it has run out of humus,' he said. We agreed.

From the book of the Kinship, I shall only cite Howard Jones, a leading market gardener, who mentioned rotations and return of the humus, and Lord Portsmouth's summing up which will be the motto of this present chapter: 'In farming especially, cash return itself rarely represents true values.'

Then there was, and still is, the practical work of Lady Eve Balfour who must rank as one of the most important agents for truth in the debate. She set up a trio of experimental farms which have been continued by the Soil Association and are open to inspection. Many results can be seen but this account will be confined to the main points only.

During the trials, which lasted for twenty-three years, the smallest of the three farms raised cereals and cash crops, without stock, but using adequate amounts of artificial manures and sprays. This was typical practice on many profitable farms in West Suffolk, and for the sake of brevity I shall dismiss it by referring only to the exhibit of a core of soil and subsoil that resulted from this treatment. Any farmer who saw this core would, I am sure, be confident that he could do better by using the land in some other way. The layer of live soil on the top is hardly noticeable alongside the top layers of the cores from the other two farms. One accepts, of course, that they are typical, fair samples.

Turning to the two farms that provided the main comparison, these were each of seventy-five acres and were matched as carefully as possible at the beginning, and managed afterwards as scientifically as the staff could arrange. The rotations were of ley farms with six years in crops and four years in grass. They were self-contained, and primarily produced milk, for which the two herds of Guernsey cows were also very carefully matched. One farm used 'organic' farming methods only; the other used muck from the cows plus chemicals and sprays, and is therefore conveniently called 'mixed'. The organic farm composted its manure; the mixed one did not.

Taking the average yield of milk per cow as the integration of results of each farm, the answer came that the organic farm, receiving as we have seen, none of the artificial manures nor sprays, did about ten to fifteen per cent better than the other, according to the *Haughley Research Report*, 1938–62.

I discussed in correspondence with Mr. Robert Waller of Haughley, the performance of the two farms which I visited in 1970. He told me that the men who worked the mixed farm believed that it had given the poorer results because the bought fertilizers upset the species of grass during the grass sequence in the alternate husbandry. The differences in average milk yields between the two farms over the years are not very great but they seem well established: for example there is a figure of average milk per cow of 6520 gallons on the organic farm in 1969, compared with 5383 gallons from the mixed farm.

There are people who consider 'bag muck' to be poisonous, but it seems to me to be difficult to support them because plants can be made to grow on it provided that it is dilute enough. Nevertheless, bag muck can do harm through the ecology. Plants which are able to obtain the benefits of the decayed humus will naturally take only what nutriment they need; the bag muck contains essentially the same salts but the composition will inevitably be subject to human error in deciding upon the best constituents. Another very important point to be considered is the activity of the humus-consumers—wireworms, fungi, actinomycetes and the rest. Recognition of these two points seems to me to push the debate over to the organic side: in practice compost beats not only bag muck but also careless muck spreading. It is not the chemical that is needed but the turning of the old heap; probably this is why Lady Eve's compost-fed farm did better than a mere 'muck-fed' farm.

I judge that there were three lasting effects of the activity of the Kinship and of Lady Eve. One was the founding of the Soil Association, in which very many thoughtful people all over England kept alive the idea that right farming is organic farming. As well as continuing Lady Eve's comparisons of adjacent organic and 'modern' farms, the Association's farmer members elsewhere demonstrated that conservation farmers could hold their own commercially over the years.

The second effect was an important one: the revival of the word husbandry, which recalled an honourable past. That revival survived even into official names and the word could not help but convey something different from exploitation.

The third effect was the practice of compost-making in gardens, which is widespread. In the absence of a source of farm manure, there was many a gardener and allotment holder who grew crops successfully with a compost of vegetable waste—activated by a small quantity of dung if he could get it.

However, as our talk in Adrian Bell's field showed, it was not a time when much that was tangible could be achieved, although we could enthusiastically support Sir R. G. Stapledon's plough-up policy and ley farming. We did receive some lip service. The organic fertility of the soil was politely mentioned in the Farmer's Club and official circles, but was mocked in the smoking rooms and corridors. We were called the 'muck and magic' school. Nevertheless, for a decade or two some of us thought that we had made an indelible mark, that although fertilizers would be used, they would be used only as a supplement, and there would never be another dustbowl made in any educated country, nor land run down as Adrian Bell had found it on the farm that he took over.

The controversy decided: main facts

It came as a severe shock to me in 1966 when I was unable to sell steers at Preston for appreciably more than I had paid for extra calves in the spring. It was a still greater shock to discover the explanation, namely that farmers in the Midlands and South were no longer wintering beef cattle. Only the Yorkshire men, I was told, were still doing so, and that caused the price to recover a little later. Those closer to farming fashions than I must have known about it sooner.

The fullness of what had happened was recorded for all time in a source that I had kept aware of over the years, namely the *Journal of the Farmers' Club,* representing the very best information and criticism in agricultural circles. From it we may take as historical and authentic the revelations of 1967, which date is thereby to rank in farming biology with 1840, the date of Liebig's report to the British Association. The name for 1967 is Henry Fell. Here is macro-ecology indeed: the final evidence ecology needs to render it fully satisfying.

The chairman's introduction revealed Henry Fell as an agricultural scientist who gained a Gold Medal in 1947 and went into farm management. He was evidently no mere scientist because he rose to be managing director of a company farming 2150 acres of rather heavy land in Lincolnshire, mainly for corn.

In order to appreciate Fell's paper 'The Deterioration in Husbandry Standards', it is necessary to know where the source of profit lay in the farming of the 19th century, represented by the 'four course shift'. In the lowlands in the last century grain formed the major or the only profitable product for sale, what is called the cash crop. The farms also carried animals, perhaps bullocks or sheep, to consume straw and root crops, turning them into meat and manure for little visible cash profit, if any, but equally for not much loss. The land was scrupulously weeded during its turn in roots, which were grown in rows. The soil was refurnished with humus by its two years in grass and clover mixture plus the yard manure which was carted on and spread. From about 1920 onwards there came a modification. Corn farming looked towards two other sources of profit. One was milk instead of the non-profit meat, another was sugar beet.

But the changes of the 1920s could not be said to have destroyed the high standard of the typical husbandry, from which Fell recorded that standards had declined. He stated that from 1957–67, steadily increasing economic pressure had been put on the farming community. Capital needs were rising, 'almost every item of cost rising, the return of virtually every product falling'. Under this sort of pressure each item in husbandry was scrutinized in order to make it profitable. All were to be, if not cash crops, at least not visible loss crops, so he said the 'traditional livestock of the arable areas are steadily disappearing'.

Fell instanced Lincolnshire which reduced its sheep population by as much as 10 per cent in 1966, 'and of course, at the same time the leys go to the plough'. Even in traditional stock rearing districts 'barley is pushing sheep and cattle on to the poorer and cheaper land'. He instances the change observable as one travels through the Scottish borders. All this had been in progress since about 1952; 'the facts are not in dispute'. In the extreme cases, barley was grown year after year on the same land. This gave the farmer a welcome rest in winter. I have read a saying quoted by Mr. Tristram Beresford as, 'six months in barley and six months in the Bahamas'.

To return to Henry Fell's paper, effects were visible in three directions: in the increase of diseases, in weeds, and in shortage of manpower for cultivation. It will be realized that farmers keeping livestock have had men to do the winter tending of the animals who were free to work on the fields in summer when the beasts were at grass. On weeds, Fell saw the solution in terms of the judicious combination of herbicides and cultivations, spraying alone providing no cure.

Mr. Fell was followed by Mr. T. Cave, a large-scale farmer from Wiltshire with successful experience also in Wales and Cornwall. Kale and root crops, on which the cattle feed, declined in Wiltshire by more than 50 per cent between 1960 and 1966, and some farmers were running up to 80 per cent of their acreage on cereals. Cave said that Imperial Chemicals and Fisons were trying to foster beef and sheep, and pigs, which last can also fit on a grass break or ley. In the discussion Fell suggested: 'Why not put the money in to meat production and let the farmers sort it out for themselves?' Colonel Houghton Brown, the chairman, described Fell's paper as 'like an arrow in the hearts of a good many of us' and at the end of the meeting said that he had been watching in the gallery the ghost of the late Sir George Stapledon, with a benign smile on his face saying, 'I told you so.'

Thus Henry Fell's paper brought in macro-ecology and answered the great question: can we farm without animals? The answer seems to be clearly that we can not, for even a short run of years shows that the crops become unhealthy. I have heard of men who have tried it by adequate green manuring with rotation of clover; but have not heard whether that has proved good over the years, which I think I would have done if it had served well.

The interest of the Farmer's Club in the great question was sustained. In the issue of the *Journal* of October 1968 entitled 'Soil: are we developing fertility', part I, the subject was taken up again by two more experts B. S. Furneaux and G. H. Brenchley, both of them, like Fell, having transcended their purely technical qualifications. The November issue continued the discussion with Part II, a paper by R. Percy.

Mr. B. S. Furneaux was trained in 1927, had been on the staff of Wye Agricultural College and had then set up as an independent consultant and soil surveyor. The interest of his paper is largely on the roots and the organisms in the soil, as has been described in previous chapters of this book. He is to be commended in that he is not afraid to confess our great and general ignorance of the soil, thus 'when we are putting nitrogen on . . . there will be an effect on the soil population. One form of nitrogen will encourage one set of bacteria and another will encourage another set, that is where we are so desperately ignorant.'

For the sake of soil structure, Mr. Furneaux stated that 'livestock farming and arable farming must go absolutely hand in hand over a large part of Britain'. He wished to see straw put back into the soil. He said that the great contribution of leys generally comes only in the third year of laying down. He was anxious about the damage done through over-consolidation by heavy machinery. However, Mr. R. Percy, a versatile and successful farmer in several countries, thought that he overcomes this risk of consolidation by using very high power wheel tractors which go too fast to crush the land effectively, and have wheel tracks fourteen inches wide. The wheels run in the furrow of every fifth row in four-share ploughing but Mr. Percy had never seen any ill effects in that row. That seems to be a decisive observation.

Mr. Furneaux's paper thus serves to introduce the synecology of soil population and it may be useful to consider how we stood at this point. Mr. Fell's paper had shown without doubt that departure from the rules of the four-course shift had been followed by deterioration. There was no doubt of the historical facts, whatever the explanation. The explanation was not proven. The trouble might, for all that we have written so far in this chapter, have been sunspots. Mr. Furneaux's paper lent the weight of experienced opinion to the suspicion that the fault lay in neglect of ecology, but that is merely an opinion until it can be shown, from ecology, how the deterioration would naturally follow the new practices. To do that, however, we shall have to enquire into the history more closely, seeking more technical detail on the deterioration, something more specific than the vague 'the land needed rest'.

Particular facts

For the natural explanation we may turn to the paper by Mr. G. H. Brenchley. He had recently been Regional Plant Pathologist, in the National Agricultural Advisory Service for the eastern and midland areas.

As everyone is aware, there are so many thousands of scientific papers published these days that it is impossible to keep up with them and the ones of which the public hear are not necessarily the most interesting nor most useful, nor the ones that will be remembered. Mr. Brenchley's, on the other hand, should be remembered for a long time to come. It is of immense importance throughout the world, because it concerns the food supply and all the social things that go with productive, healthy land. Compared with that, much of the science that does reach the public press seems trivial.

Eighteen months earlier, Fell had opened his paper by mentioning the economic pressure to which farmers had been subjected. G. H. Brenchley repeated this in an epigrammatic way when he said that the pressure towards intensive corn growing was economic rather than agricultural; 'it is certainly not what any plant pathologist would have recommended'.

He mentioned the epidemic of the new or newly discovered race of yellow rust fungus called Race 60 which attacked the wheat Rothwell Perdix in 1966. This new wheat had formed only 1 per cent of the total acreage in 1965. In 1966, when the epidemic occurred, it was 8 per cent of the acreage and out it went. The fate of Perdix was, he mentioned, by no means unique. 'During the past twenty years, some six varieties of wheat have been virtually eliminated in one season by somewhat similar epidemics.' He names the varieties and quotes the years: 1949, 1952, 1957, 1962 and 1967.

Experience with so-called resistant varieties had not been encouraging. 'New races of the fungi causing yellow rust of wheat, barley mildew and potato blight have appeared that are capable of attacking varieties that are specially bred for resistance to these pathogens.'

Brenchley made the important generalization that the fungi were often blamed wrongly. Often he had noticed that fungus had attacked where there was bad cultivation of one kind or another; he mentioned take-all which is a soil fungus that frequently attacks cereals and may cause the roots to go to a spongy mess. The fungus is there all the time, 'waiting for the farmer to put a foot wrong'.

The passage on remedies should have shocked many of the agricultural correspondents of the day. There was much written on fungicides and many fungicides were sold. Brenchley said that they were troublesome and expensive, and, worse than that, they were effective only when the fungal spores are few, that is, if they catch the disease at an early stage. Even then, their principal usefulness must be in air-borne diseases. They are of doubtful value when the disease is soil-borne, as is take-all for example. The final recommendation is control by cultivations, rotations and good drainage and good soil structure. Brenchley reserved the use of chemicals and fungicides for particular operations, for example the destruction of self-set seedlings of wheat, which may carry an infection over from one year to another.

Thus, in general, modern farming and modern aids are condemned and the panaceas are denounced. Even the break-crop of oil seeds etc., which was being officially advocated, seemed unlikely to be of any value because a break of only one year, even it it were grass itself, did not do enough good. Thus, finding data detailed in Brenchley's paper, we know that one considerable symptom of the deterioration was, as Fell has said, the onset of diseases. Brenchley's paper shows the diseases beating all modern and expensive remedies.

Those were the facts, needing no ecology to establish them. It was still, however, arguable that the triumph of diseases was due to some other cause than modern neglect of humus, as Brenchley suggested. To pin the matter down requires an explanation in terms of the synecology of Chapter 3.

Let us then begin with a simple case, which none would doubt: the spoiling of a crop of potatoes by wireworms on ploughing out a pasture of ley. Under the turf the wireworms had fed unnoticed on obsolete grass roots; in the absence of these they turned to the potatoes. We may wonder whether a similar transfer of attention would occur among other soil organisms that normally live harmlessly and helpfully on humus. There is certainly a loss of humus in cultivations, as there is also, by the way, when the land is limed. The main point, however, is that on arable land every cultivation hastens the decay of humus by the natural agencies that we have detailed in the previous chapters, for example the aerobic bacteria and all the other soil organisms that receive exercise and stimulation in the processes of cultivation. Nevertheless, the use of artificials enables the farmer to obtain a crop in spite of the fact that there is less humus in the soil than there would be if he were relying on the return of dung from animals. So, the farmer puts a crop there, of which the roots have to contend with all those organisms that have now less than their former supply of humus on which to feed. As in the case of the wireworms and the potato crop, there are nematode worms, mites, fungi, actinomycetes and bacteria, to say nothing of other organisms normally involved in the process of decay of vegetable remains, all probably capable of reproducing as fast as formerly or nearly, and now no longer provided with their natural food, obtainable formerly from decay of vegetable remains. In those groups named, it is known that there are some species able to live entirely free in the soil on what they obtain from the air; others live wholly as parasites on plant roots and have no other way of living. Between those two extremes there is, in every group mentioned above, a vast number of species or races which live mainly on the breakdown of humus itself. Bearing in mind that there are always plant roots in an obsolete state ready for breaking down, it is not to be doubted that there are organisms which can switch from eating dead and decayed roots or manure to attacking live roots. Often only slight damage to the plant skin will allow a parasite to enter. Nor is it safe to assume that existing species are the only ones to be feared. New strains have appeared, and it is almost certain that every effort of the plant breeder to obtain immune varieties will produce a corresponding strain of fungus or bacterium to attack it. Some bacteria are capable of producing three generations in an hour, whereas a new crop strain cannot be developed in under a year, usually more. If there is a race in breeding new strains it is not difficult to foretell which group will win it. To put the crop plant into the soil where the organisms that are normally engaged in the decay of humus have not sufficient humus to occupy them, is to invite these organisms to attack the roots of the new crop plant. That is sufficient explanation of what has happened to farming in recent years and there is nothing vague about the chain of consequences.

Even dustbowls

News of the trouble even reached *Drive*, a motorists' magazine. Mr. Jack Lowe, the County Planning Officer for Nottinghamshire, reported that in March 1968 he looked out from his house and saw 'nothing but rolling clouds of dust filling the sky'. The same event was reported and photographed (Fig. 13) by Mr. D. H. Robinson, in the *East Midland Geographer*. It occurred chiefly because several thousands of miles of hedges have been uprooted by farmers and contractors, with government assistance—some say 7000 miles, some 10,000. And the soil has lost the adhesive value of humus. Even the loss of hedges is correlated with the loss of humus. Mr. T. Cave reported that in Wiltshire there was an average decline in kale and roots of 50 per cent of the acreage between 1960 and 1966, whereas corn had increased by 50 per cent. He nevertheless sees some hope of farming crawling back by the use of bullocks and sheep. These animals would require men to look after them, such men as had been available for the care of hedges during the summer. Part of the incentive for uprooting hedges has been that there is less labour to keep them in order. When men leave the land it is difficult to get them back again. It may seem therefore at first sight that Mr. Cave's suggested method of recovery is out of reach, but probably men could be recruited from the poverty-stricken family farms in the more mountainous districts of the North and West of Great Britain, where for generations good south country farmers have been recruited.

In January 1969, Mr. Jack Meyricks, the Chairman of the Farmer's Club, was reported as recommending the return of three million acres of worn out cereal land to grazing, 'preferably to sheep', as a measure to 'go some way' towards restoring the fertility of land in this country. Much land formerly would grow good crops of wheat, but 'will not now even grow a reasonable crop of barley'. It is possible that the Chairman's weighty opinion reversed government policy, like a death-bed repentance, but not so final

13 **Wind erosion on light soil.** Dustbowls are rare in the British Isles, but D. H. Robinson of Louth who described the events of March 1968 in the *East Midland Geographer*, has provided this photograph from his survey, and the two paragraphs which follows. Those are his words, not mine. I would be tempted to be less tolerant of what I regard as short-sighted policy but I regard his words as sacrosanct, being as near to the event as any of us are likely to reach. He writes:

March 1968 was a dry spring, following a dry winter with more frost than usual, which coincided with a period of strong winds. In Eastern England in particular the result was severe soil erosion by blowing. Light soil lands, for example, those on the Spilsby sandstone at the southern end of the Lincolnshire Wolds suffered badly as this picture shows. In places the landscape was obliterated by clouds of blowing soil.

In the late 1950s and in the 1960s agriculture in lowland England took a dramatic swing towards intensive arable cultivation. This involved larger fields for larger machines and more economic working, hence the amalgamation of small fields and removal of their hedges. While hedge removal is not a primal cause of soil erosion by blowing, it has certainly been an aggravating factor. There is also evidence that soil structure has changed because of reliance on chemical fertilization rather than manure, and hence light soils are more susceptive to blowing.

because anything to do with government is likely to wobble if it proves unpopular. The evidence of a new grace in the principal harmful agency was in *The Times,* 26 May 1969, 'An agricultural correspondent' wrote that, 'as recently as last November there was strong official talk of concentrating sheep development away from the lowlands', but all over the country, in 1969, experts advocated lowland sheep. The Minister said, 'Sheep are important to the husbandry of many lowland farmers.' It is known, however, that the restoration of fertility in run-down land is an expensive and tedious process.

The history, from 1840 to 1967, should be a warning to scholars to be especially cautious and doubtful when their own living or importance is tied up with the soundness of a discovery, as very often it almost must be. Towards the end of the history another warning becomes topical: a caution to indoor men—in this instance economists—against propounding panaceas for outdoor activities, where life is often beyond their comprehension. The proof that plants could be reared on minerals alone in water culture, was presumed to apply in soil.

Although it is cold comfort to note it, the blunderers provided a study in ecology without an equal, at all three levels, and we can state for farming with a certainty that we can not match for any other eco-system, that the determinant is the cycle of humus. We do not need to add the phrase, without which the eco-system could possibly, we think, collapse, because we can state, as a fact, that without the cycle and renewal of humus it did collapse, over a large area of England in the period 1952 to 1967.

It will be interesting to see what features in other eco-systems resemble those of the farm, where after all there is a certain universality, even to the extent that the life processes take place in an aquatic medium—namely the soil solution.

5 Aquatic ecology

Ponds

Among the fields of the farm we find the ponds and streams and by now we can expect that life in the waters will not be essentially different from that in the soil solution of the fields.

Nor is it. But when we take up the subject on the framework we are using, that is, first autecology, then synecology, then macroecology, we find that only some of the aquatic specimens look similar to corresponding forms of the land and others are markedly different. Thus, the water-vole seems only a larger kind of field-vole, and both live on vegetation; the water-shrew, feeding on insects, is very like the common shrew of the same feeding habit. Whereas when we turn to the birds we find some, such as herons, kingfishers or ducks and grebes, so aquatic that they have no obvious relatives living on the land in Britain. Amphibia, that is frogs, toads and newts are firmly of both land and water, as named. None of the fish has anything to do with the land, except the eel which travels overland in wet grass in the night. The eel is in several ways an uncommitted animal, undoubtedly a fish but pre-eminently an eel, and to the minds of some simpler people as strange as a snake.

Turning to the smaller animals, the beetles are but little modified for aquatic life; whereas water-boatmen and water scorpions and pond skaters are quite peculiar, and dragonflies in the juvenile form, have no land relatives like them. Water snails look and behave like land snails, but the related mussels are aquatic mystery purses, which may contain treasure. Mussels live on grains of organic matter, which they filter out of the bed of the pond.

Admittedly, the aquatic forms that closely resemble terrestrial are few, but that any can be nearly the same is sufficient to make it seem likely that the life of the water is basically similar to that of the land.

The plants provide indications resembling those of the animals, the sweet grass *Glyceria aquatica*, which blocks the disused canal at Salford, is very similar in size and shape to cocksfoot grass, but the pondweeds do not live on land. The same applies to herbs of various groups, where there are genera that grow only in water: water buttercup, water dropwort, water lilies and others.

When we look deeply into the water of a clear pond we may see living specks making curved trajectories of six or eight inches, in this direction or that. These are crustacea of the group *cladocera*. Others also move intermittently, but in straight trajectories and shorter, these are *copepoda*, and there are similar less mobile forms belonging to the *ostracoda*. Those and any other animals that live mainly in the body of the water, though they may settle for a while, are called zooplankton and the microscopic plants on which they feed are phytoplankton, of which the members that settle for greater or longer periods on pondweeds or on mud or on submerged objects constitute the periphyton.

Periphyton is so obvious in the pond that it is naturally, and probably correctly, taken to be of importance in the chain of nourishment, and it is the seat of the greatest resemblance between soil solution and pond. In the phytoplankton generally, there are the same classes of plants as in the soil solution, diatoms, green algae and blue-green; and when these are situated in the periphyton they demonstrate their close resemblance to the kinds in the soil solution, often even identity.

Autecology in ponds is so well known that it is also possible to lay out some of the synecology (Fig. 14).

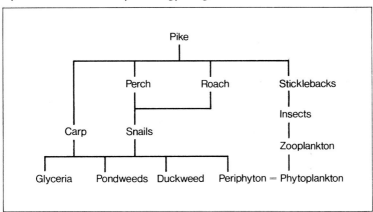

14 Synecology of ponds; plankton and pond weeds. Two pieces of web are shown: that involving plankton, and that from pond weeds. The basis on the ground floor runs from *Glyceria* to the phytoplankton with a bridge or passage in the periphyton connecting the two wings of the structure above. Both stairways lead to the ruling fish, the pike. There are many missing elements in my diagram, both of species and of connections, but the 'wheels within wheels' world is very clearly indicated, by even this short version, up two stairways.

All the animals shown so far feed and excrete, producing the detritus, as their rubbish is called; it consists of fragments and micro-organisms. The detritus mainly falls on the mud or is caught among the periphyton, and we can make another diagram (Fig. 15) for what happens to it. The diagram is to be read from the bottom upward.

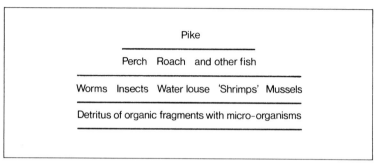

15 Synecology of ponds: the detritus. This evidently forms a basis for a third wing or staircase to the pike, nourishing a rich fauna on the way.

As we consider Figs 14 and 15, we cannot help but be impressed by the part played in the first by the pondweed. Growing in the mud, it provides shelter for numerous organisms of many kinds, food for others, especially the ubiquitous snails, and is the main source of nourishing detritus, which supports so many more kinds of animals. Duckweed and *Glyceria* doubtless have important roles but they may be absent, whereas most ponds have at least a little pondweed and many have much, growing up from the mud in the bottom of the pond. This important part of the synecology grows from the mud, which therefore seems at first sight to be the key to life in the pond—the determinant—mud.

That is to say, that a flourishing pond life depends primarily on flourishing mud, which doubtless behaves as in Pearsall's rice fields, in respect of pH and more especially of redox. In all the variety of ponds the virtue of mud is likely to be the dominant factor, by the pond's mud the fauna and flora will live or die. In ponds where there is little mud, there will be little biota, that is flora and fauna.

It is therefore all the more remarkable that if one looks at the case of the pond on another time-scale, mud is not the determinant but the hazard to the pond's existence—the anti-determinant one might say.

I look after two ponds. One is a small round pond of some 20 feet diameter. It appears to have neither inlet nor outlet, but there must be some means of seeping away, because when sealed with an ice cover the water falls away beneath, leaving a cover hollowed by its weight. And normally there must be small springs or seepage to feed the pond, which in eight years has never been dry. There are grasses growing into the pond from whose stems the caddis make cases, and it is the resort of many newts. Under a post on the bank I found a nest of caddis cases, which I thought to be a water-shrew's. There is very little pondweed if any, but some *Glyceria fluitans* and seasonally a nice cover of duckweed.

A larger pond is about 100 feet long by 20 feet across and a current flows through it—yet it continually threatens to silt up and become a marsh filled with sedges and sallows. After a heavy rainstorm it is ochre-coloured with silt from a roadway on higher land from which it takes the drainage. In the meantime it is richer than the other, with voles and frogs, toads and a wealth of sticklebacks. However, were it not for periodical digging, this pond would cease to be a pond, and the original work that dug it, for the sake of sand I suppose, would eventually be obliterated, as the sallow 'carr' became more and more solid, growing alders, then sycamores.

Comparison of the behaviour of the two ponds is a study in macro-ecology. The one is self-maintaining, the other is self-obliterating. The nature of the water each receives enables us to state a new determinant for the pond, namely a supply of clean water. So we have an instance slightly more complex than that of one simple determinant: we have that the pond-life is determined by the mud; but the pond's existence is determined by a supply of clean water.

Dr. Popham agrees with me in thinking that the ecology of our local canal in Salford is very like that of a pond. The biota is in general similar to that shown in the diagrams above, and a canal in use is kept open by a combination of maintenance weed-cutting and a supply of clean water from springs. Although little else is added to the canal save dust, the productivity may be quite high. A method devised in the Biological Department yielded a first estimate of 5000 roach, of a length of 8–12 inches, in a 380 yard stretch of the local disused canal.

Streams

A sluggish stream is not very different from a pond that has an appreciable current in it, but when the current is faster we have apparently clearer water and a cleaner bed with little or no mud or soil above the stones or gravel. At high speeds the water-surface breaks on obstructions and so may absorb air. The faster the current the coarser the particles the stream will carry. The fish will be species of *salmonids* (game fish) instead of predominantly *cyprinids* (coarse fish). The main food for fish will be sessile caddis but other fauna will lodge in crevices and hold on. They will even hold on behind waterfalls, as Dr. Popham has described in *Aspects of Life in Fresh Water*.

Probably none will disagree if I state the determinant of life in the stream to be the speed of current. Altering the speed of the current will alter the character of the stream by means of a series of predictable stages.

Lakes

Round its margins a lake shows many features already seen in ponds, but the bulk of water is so very much greater as to make differences in kind. Looking back at the web of life in the pond, the wing or staircase leading through the plankton is the one we can expect to see developed; and so it is, but the great size of the lake makes it a new environment.

In particular, the elements of wind and water are massive enough to take charge at times as is rather simply illustrated by John Murray's observations in Scottish lochs. As the wind blew across the loch it drove the surface water up against the shore which induced a bottom current—the 'undertow' feared by bathers—so that the whole loch was in circulation, a current going with the wind on the surface and a counter-current in the opposite direction at the bottom.

At some intermediate layer the water is being pulled both ways and at least small vertical motions are set up, which have only to become large and strong as the wind freshens for the whole body of water to become totally mixed. In a deep lake the lower layers may not be affected in this way.

Lakes often show a resistance to total mixture in an interesting and important way. I have extracted here an explanation, and partly their diagram (Fig. 16), from *Life in Lakes and Rivers* by T. T. Macan and E. B. Worthington.

First they consider the winter temperature and the depth, and find the whole body of water to be at 4°C from surface to bottom, evidently well stirred by strong winds.

From May onwards, the surface water is warmed by the sun and because its density is thus reduced, the same water tends to stay at the top. The gentler summer winds may disturb it, but not to any considerable depth, and not strongly enough to change the system of lighter water over heavier. Nevertheless, the wind disturbs the top layer and by July the top layer of 10 metres thick of warm water—at 15°C—has separated from the main body, which is at

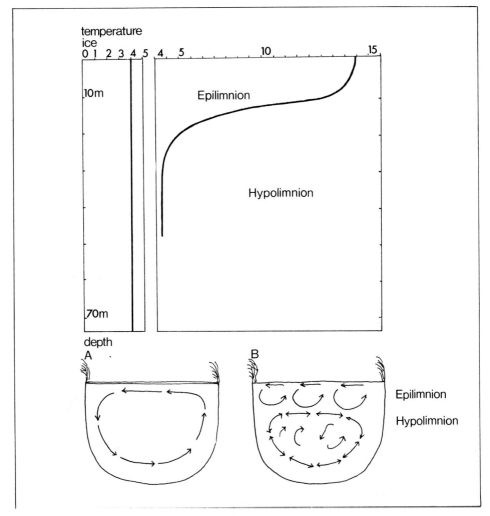

16 Thermocline and seasons in a lake. The upper diagrams show temperature plotted against depth on two occasions in Windermere. The lower are models to illustrate the following: (A) When fresh water is cooled to 4°C it sinks; when cooled further it expands a little and so stays up. When cooled to 0°C it freezes and floats. Before the protective ice cover forms completely, the wind can circulate the water as one body, from top to bottom. (B) The summer sun warms the upper layers which, being thus made lighter, do not sink. The gentle winds stir them a little to a moderate depth; but do not disturb the lower water: so forms the thermocline. The thermocline is not broken until autumn brings gales. After Macan and Worthington, *Life in Lakes and Rivers*.

about 6°C. In this example, the transition takes place between depths of 10 and 22 metres which is said to be the thermocline. Often a thermocline is more sharply defined, perhaps being only three metres thick.

So, we have the lake in two layers, the warm well-lit epilimnion with its own currents, counter-currents and turbulence, contrasted with cool hypolimnion, a dim-lit layer below, perhaps borrowing some disturbance at the layer of the thermocline, but on the whole undisturbed compared with the epilimnion.

Below the thermocline there is not sufficient warmth for photo-

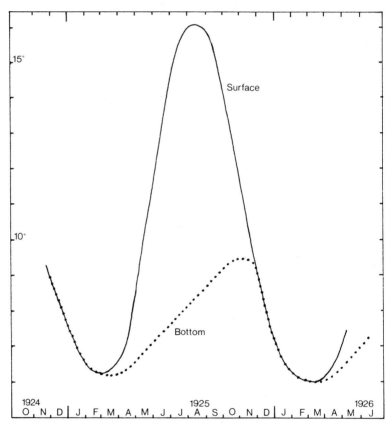

17 *Seasons under a thermocline in the North Sea.* The surface temperature is shown by the thin line; the temperature at the bottom (40 fathoms) is indicated by the dotted line. On the surface, the maximum temperature was in August, but on the bottom it was in November, after which the thermocline was broken up by gales. In shallower areas, where there was no thermocline, cod grew to 30 cm length in their first two years, but here they took three years to achieve that length. Fuller data in Graham (1929).

synthesis of green cells to compensate for the loss of oxygen due to decay of corpses falling from above: thus plant nutrients are unused and accumulate in the hypolimnion.

Above the thermocline in the warmth and light, microscopic plants can flourish, but by July the outburst of diatoms, which takes place in spring, is often over. By this time there is usually a decrease in nutrient salts, such as phosphate, as well. These two phenomena are generally thought to be effect and cause. There is another outburst of diatoms in the autumn, which, according to that school of thought, is caused because autumn gales stir up the nutrients which had, during the summer, accumulated in the hypolimnion.

Sometimes there are other explanations in particular cases. Diatoms, as they reproduce themselves by binary fission, become continuously smaller—the older the smaller—which cannot well go on indefinitely. However, the process is corrected from time to time, when the diatom cells form larger organisms called auxospores which rest awhile, so that the increase in numbers and decrease of size has a check. Furthermore, the spring outburst of diatoms is followed by one of copepods, which graze them down, and from which the faecal pellets need a little time to rot before they become nutrient salts and food for diatoms again. Thus the fluctuation in diatom numbers might be due to predation by copepods.

Before leaving the subject of lakes, we should note the late Professor Pearsall's classical observation on their differences due to geological age and to differences in the hardness of the rock of the lake basin. They differ from cloudy to clear, as streams do from coarse fish to salmonids and coregonids, from many species to few—and almost as the two ponds I mentioned. The lakes of the north-west form an instructive series, we might say, from Wastwater to Esthwaite. Part of Pearsall's theory is that lakes tend to become gradually more and more productive through a saving, as it were, of organic matter, and that if one wants—as water companies do—to maintain the clear water type of lake, it requires hard fishing, and thus efficient removal of organic matter at the stage of salmonid and coregonid species.

The thermocline and its profound effects on the ecology of lakes has been noticed only during the present century, and it may be early to give it the importance for which it seems to qualify. The productivity of a lake and the incidence of seasonal changes, in the lakes we know well, seem clearly to depend on whether the lake commonly forms a summer thermocline or not. Great lakes are too peculiar to be lumped with the general multitude, and there are many special lakes, salt, alkaline, tropical or ice, but on the whole, in temperate latitudes at the least, the factor most strongly competing to be recognized as determinant is the thermocline.

Estuaries

Dr. Popham has studied the Ribble estuary, comparing the fauna with that of some other estuaries of which he published an account in *Oikos* (1966). He noticed effects of traces of fresh water, but suggested that particle size in the substrate is the determinant: there were very few species flourishing in mud. Dr. Popham's suggestion connects with findings off shore in the Dogger Bank area of the North Sea and with what we have found in soil and fresh waters about the value of aeration.

The rocky sea shore

It is peculiar that such a narrow and extended habitat as a shore can be considered as eco-system, and certainly few stretches of shore could be expected to be even as much a closed system as a lake. All eco-systems are in fact open systems, but the seashore looks more open than most. Nevertheless, it has its more or less stable characteristics. A. J. Southward's *Life on the Seashore* is an excellent book about this interesting but peculiar eco-system. A population of animals and plants clings precariously to the intertidal zone, and the fauna and flora is especially rich where the layman might think that conditions would be hardest—on rocky shores, in the pools and crannies. Again we seem to see the importance of a firm substrate and good aeration.

On arrival at the seashore, it is obvious that we have something drastically different from the lakes and other fresh waters. The larger plants are brown, having a pigment extra to green chlorophyll, and they grow on the substrate not in it. There being no stable soil, seeds could not survive to take hold. The seaweeds liberate enormous numbers of actively swimming spores, of which a successful one will have managed to attach to rock or stone and straightway grown its 'holdfast', in what cranny or roughness we do not well know. The new growth, following closely the form of a substrate, has a firm hold, indeed a remarkably solid hold, which cannot be pulled loose, even when the great frond grows and strains on the stipe or stalk. Thus the plant is made for a world of turbulence and aeration. We may expect that sea-water of rocky shores will be high on the redox scale.

As well as brown seaweeds, there are some red ones, and a few only are green. The red pigment is supplementary to the chlorophyll of green plants, efficient at low illuminations.

Returning to the shore, one of the most noticeable points is that so many of the organisms are hard-shelled or spiny or thick-shelled. They seen to have abundant lime and to spare, whereas on land lime seems nearly always to be in short supply, except over rocks that were formerly under seas. Like blood, sea water is slightly alkaline: we are in the high numbers of pH as well as in the high redox. The combination is doubtless what makes sea water such a good cleansing medium for the countless tons of horrifying wastes discharged into it, both domestic and industrial.

The most striking feature of the rocky shore is zonation with tidal level. The type of fauna and flora can be quite accurately predicted according to the tidal levels. Examples are afforded by brown seaweeds of the common genus *Fucus*, the flat, bladder and

18 **Limpet the great grazer.** Limpets were removed from a strip 5×110 metres by N. S. Jones (1948). The strip first grew green algae in a thick felt, and later the big brown seaweeds. The very numerous limpets normally there evidently live on minute sporlings of the algae, thus keeping the prey invisible but none the less important to the predator. (The photograph of which this is a detail was originally published by permission of the Air Ministry with all rights reserved.)

knotted wracks. However, although naturalists have studied the inter-tidal zone for nearly two hundred years, they have as yet no certain knowledge of the factors which cause such zonations. Some shore organisms can withstand more exposure to air, or wind or sun than others: that is about the sum total of our knowledge.

When we consider exposure to the waves, the story is simpler and better known.

As a student, I compared the relative height of shells of limpets in exposed and sheltered sites and found significant differences, which have since been published and confirmed many times. A limpet fits against the rock exactly, so exercising a hold by suction, as well as reducing desiccation. After grazing it homes to its own site. Limpets are responsible for the bareness of many rocks in the inter-tidal zone; N. S. Jones from Port Erin has shown this at at Port St. Mary nearby, by keeping an area clear of limpets and so allowing complete cover of seaweed, with a striking air photograph (Fig. 18).

In my experiments long ago, I was interested to know whether a limpet went straight out and then straight back along a single known path. By smearing a circle of grease an inch wide around a few limpets and inspecting them again on the next low water, it was possible to see that they had often returned to site by a route other than that which they took on leaving. Its own part of the tempestuous and open habitat of the shore dwellers is evidently a well-known home to each limpet.

Exposure whether to air for part of the tidal cycle, or to the battering of seas, must determine life on the seashore, where animal and plants meet the condition by an array of devices in structure and habit. Between tide-marks, exposure is the determinant.

Nearby seas

On rocks at and immediately below low-water mark of spring tides we will often find the great oar-weed *Laminaria* with its strong stipe and long broad frond enabling it to use what light reaches its relatively deep zone. If one peers down into a calm spot of that laminarian zone, or perhaps better finds a pool of corresponding tidal level, one can achieve some idea of life in the seas immediately deeper and to seaward of it. There no seaweed grows. In the sand bed of the pool there are the slight signs of burrowing worms below; on the sand surface the occasional disturbance of passing shrimp or louse; the sortie of crab or lobster from ambush, the camouflaged fish, hitherto unseen, revealing itself by a sudden dart. On the floor or clinging to the rocks are the slow-moving

Aquatic ecology 71

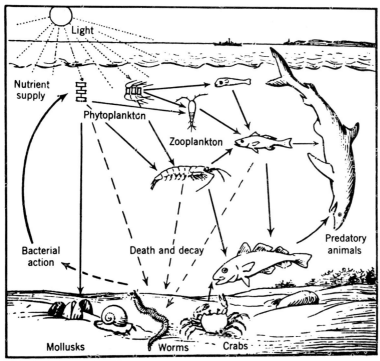

19 Clarke's web diagram. The diagram illustrates the synecology of coastal seas. The soil harbours many species of burrowers and tillers, and nourishes a few, but not to the degree shown in the corresponding illustration for land, Fig. 11 Sunlit water is of more importance here than the soil under it. Sunlight energizes the phytoplankton, which corresponds to the terrestrial grass, and the nutrient cycle can proceed entirely in water. Crustacea in salt water correspond to insects on land. From *Elements of Ecology* (1954), by permission of the author and John Wiley and Sons.

echinoderms, that is, urchins and a great diversity of starfish. At the side 'flower' the chrysanthemum-like sea anemones—no flowers but stinging animals.

For the synecology of nearby seas, Clarke's diagram (Fig. 19) seems to be very hard to improve on. The fauna and flora, the benthos, which is the name for the animals of the sea-bed, and the sunlight—all are shown, playing their part. The only point of difficulty that the diagram seems to raise is the general one on eco-systems, about the degree to which the eco-system can be considered as self-contained. If the interchange with other systems were small, as it well may be in some areas, the diagram would be sufficient, but commonly there is a contribution from the land and another from deeper water. Either may provide a substantial quantity of nutrient salts, such as phosphate. A diagram of the North Sea (Fig. 20) shows the phosphate supply from under the

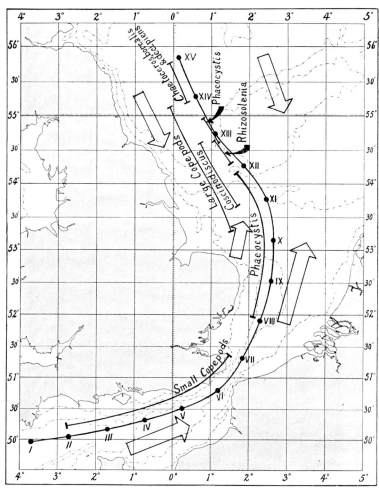

20 *Tracks of cruises on the supply of nutrients in the North Sea.* In addition, there are short broad arrows indicating the general surface currents. The track chart is a mean of three cruises in the research vessel *George Bligh*, April, 1935, 1936 and 1937. Positions i-iv have been omitted. The names inscribed are of the principal plankton in catches of Hensen's net of the order of 60 meshes per inch, but which clogs quickly, the lumina, however, remaining good enough to distinguish presence from absence, as here. Types of plankton tend to be taken in expected localities, e.g. a patch of *Rhizosolenia* in an area of upwelling. But in one cool summer it was not there, although the phosphate was. But, to mention variations again, in one dull summer there was no phytoplankton outbreak, although there was ample phosphate (indicating N.P.K., etc.) Figs. 20–2 from Graham and Harding (1938), by permission.

thermocline of deeper northern water forced upwards and mixed in the banks area, between the Spit of the Dogger Bank and the Well Bank. At the south-east extremity of the section, there is sometimes another phosphate supply, from river outfalls of Thames and Maas, enriching the water entering the North Sea from the English Channel. Both sources give recognizable outbreaks of diatoms, sometimes to the embarrassment of the herring fishery because, essential as diatoms must be to the ultimate nourishment of the herring, there seems to be a danger of too much of a good thing when the water is so full of sharp siliceous shells and their contents as to clog bolting-silk tow nets and to make ropes and fishing nets slimy to the touch.

Taking account of all the known complications, and variety, it would not seem at first sight possible to state a determinant for nearby seas. However, there is a guide from macro-ecology. In all the seas near the British Isles, and Faroes, Iceland and beyond, the effect of wartime closure to fishing, or in some cases reduction only, was to allow increase of catch per unit fishing effort, compared with the years before the closure. This is shown in a rather simple chart made at the time. Fishing had evidently been holding the weight of fish stocks below what all the other elements in the

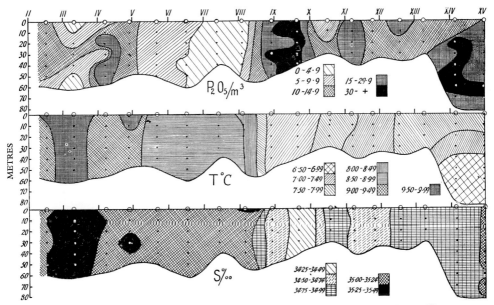

21 **The nutrient cruise in 1935.** P_2O_5 stands for phosphate, that is as pentoxide in milligrams per cubic metre. T is temperature in degrees Centigrade. $S^o/_{oo}$ stands for salinity in parts per thousand.

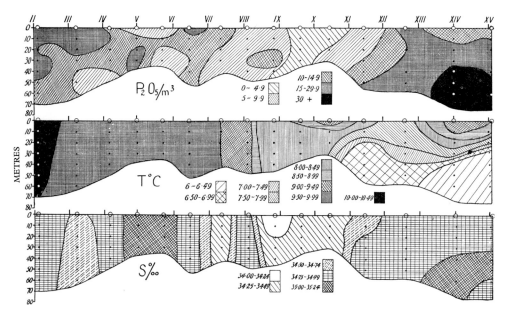

22 *The nutrient cruise in 1936.* Figs. 21 and 22 both show higher salt at the ends of the section indicating a mixture of ocean water. This is relatively warm at the south end, cool at the north, under a thermocline in 1935. The cool, deep northern water is rich in N.P.K., etc., accepting phosphate as representative of them all. Rich and relatively fresh water was found in 1936 between positions ix-x, and was ascribed to pollution from the River Thames. To sum up, the nourishment of these near waters was both from ocean and from land.

web could support, and, in doing so, must have altered the levels of all of them and their interactions. In terms of visible fish, plaice, the effect is shown in Fig. 23, and with the resumption of fishing, the effect could be reversed. We have, therefore, come to the pass when the determinant of nearby seas is fishing. Some would not think that, in such great areas, a mere scattering of fishing vessels, which rarely look many on the great expanses of seascape on which they labour, could have so great an effect: nor could they, I well believe, if the fish and they were evenly distributed, as the appearance of statistical charts would suggest. But the experience of fishing voyages always includes finding the fish. It may be in a narrow streak or on the one side only of a knoll or hole, and subsequently keeping on them. The handiest example, shown in Figs. 26 and 27 (pp. 78 and 79) comes from distant waters. There the fish lay in a narrow zone and in it only, in paying quantity. That accords with all my fishing experience, which has been more in near waters than in distant. At any time, the fish are substantially gathered only on a small fraction of the total area available to them,

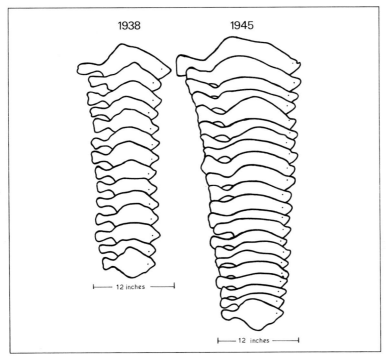

23 Mortality of plaice in near waters. Typical catches (1938 and 1945) are shown, each for fifteen minutes trawling in the North Sea, according to English statistics. Number and size of fish are both significant: wartime for men was peacetime for fish.

perhaps a fiftieth or a hundredth. In the North Sea the total area, were it fished evenly, would be trawled over once in a year, so that the fishes' haunts—a mere one-fiftieth or one-hundredth of the whole—would on my estimate receive attention 50 or 100 times per annum. Apart from the explanation of fishing controlling the numbers of fish, the facts are not in doubt. And seeing that fish are the principal predators of everything else in near waters, fishing is the determinant of the whole biota.

The ocean

Finally we turn to the ocean, the largest aquatic system. By far the greatest part of it is too far from land and markets to be fished intensively, but by the 1960s, the consideration of fishes' haunts, as just used for the North Sea, applied even in the wide oceans. In 1956 I had tried to convey my idea of the 'great matrix':

It may be rash to put any limit on the mischief of which man is capable, but it would seem that those hundred and more million cubic miles of

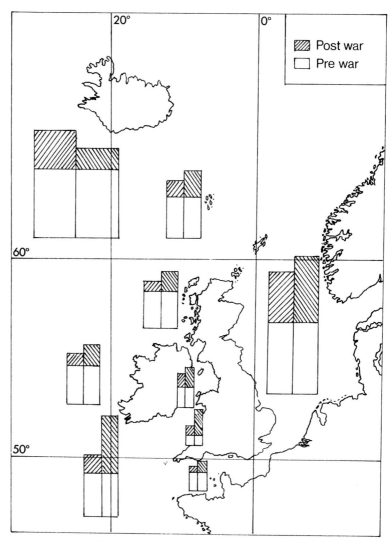

24 **Dominance of fishing in near waters.** English fishery statistics can be used to derive the bio-mass of fish stocks by the landing per day's absence of standard trawlers. There was a very great increase due to wartime reduction of fishing. This, and the preceding diagram, leave no doubt about what species is the super-predator, doubtless also affecting all echelons below these and other fish. (The left hand columns refer to the First World War and the right hand columns to the Second World War. The shaded areas show the increase in biomass in the immediate postwar years as a proportion of the biomass in the pre-war years represented by the unshaded areas.)

Aquatic ecology 77

water, containing every natural chemical element and probably every group of bacteria, supporting every phylum of animals, moving on the surface from the equator to the poles, and returning below, stirred to many fathoms' depth by the wind—it would indeed seem that here at the beginning and the end is a great matrix that man can hardly sully and cannot appreciably despoil.

I was too optimistic. Man, we know, has since sullied it: there truly exists no limit to man's mischief. Penguins have been found containing pesticide, and there are disturbing newspaper reports of pollution in low latitudes, and man has also despoiled the ocean in that the fin rorquals are now following right whale, sperm whale and blue whale to levels that will not support fisheries of former magnitude. He has also increased carbon content via the carbon dioxide of the air, from his combustion of fossil fuel. However, the description 'a great matrix' still just stands, though precariously; and if there is to be a determinant it needs to be of an enormous magnitude.

The introduction to Wimpenny's book on plankton (1966), contains a drawing of a section from pole to pole (Fig. 25), indicating the great movements of water masses, which cause nutritive up-wellings of bottom water, for, as in the lake, and for similar reasons, it is the bottom water which holds the reserve of nutrients. All that water below 1000 metres depth is many times stronger in phosphate, which is taken to indicate also nitrates and

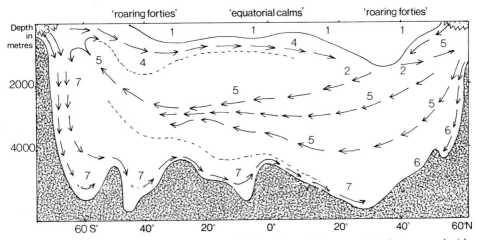

25 *Layers in the Atlantic Ocean.* This diagram is derived from Wimpenny (1966). Compared with preceding figures here, his might be called 'thermocline plus'. As shown here he recognizes (*1*) trophosphere, (*2*) Mediterranean influx, (*3*) ice, (*4*) sub-antarctic intermediate water, (*5*) North Atlantic deep water, (*6*) Sub-arctic bottom water and (*7*) Antarctic bottom water.

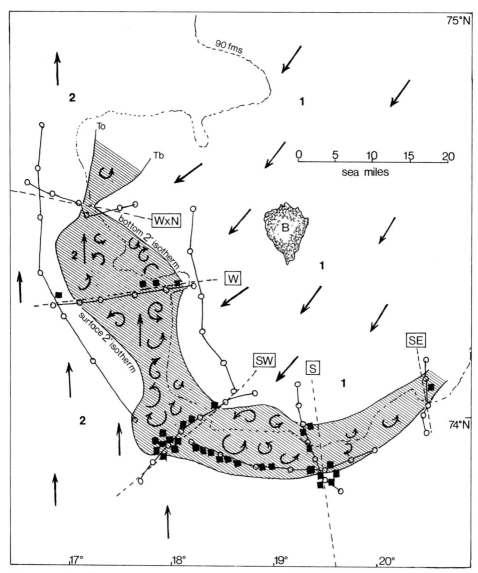

26 Contention of waters as key to location of fish. Cold water (below 2°C) leaves the polar basin and floods the area from north east to south west (1, 1, 1,). Warmer water comes as part of the North Atlantic Drift, flowing from south to north as the West Spitzbergen current (2, 2, 2,), with Coriolis force, as ever in the northern hemisphere, pressing it against Bear Island Bank. The cod fishery is shown by solid blocks in trawling tracks (indicated by connected open circles) of the research vessel *Ernest Holt*. Each block represents a 'bag' per hour, i.e. one and a half tons, lesser catches are omitted. The letter B marks Bear Island. The S.W. edge of Bear Island Bank is indicated by the 90 fms line.

27 Contention in profile (opposite). These are the data of the above figure, shown in profile. Cod seem to be held up and concentrated in the mixed waters at the margins of the warmer waters. The broken lines represent the 2° isotherm.

other nutrients, than marine plants need, and it is of enormous volume. Equatorial sun and polar ice are great factors, so are untold salt water and fresh water from ice-caps. Heat and convection doubtless account for part of the movement in the great circulation; as do winds, again as in lakes: but the one factor that is of greater magnitude than the great ocean is the rotating earth itself, giving the Coriolis force.

The rotation of the earth generates the Coriolis force in the

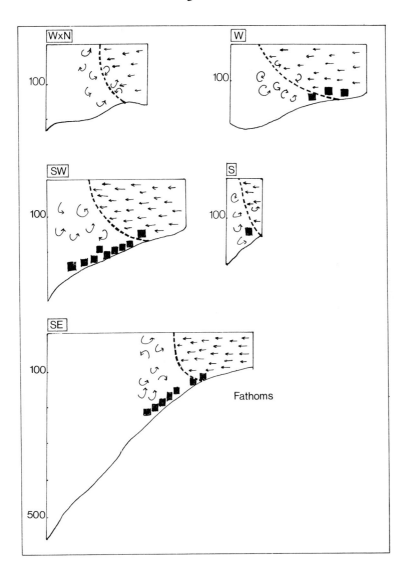

following way: if you take a globe and revolve it eastwards as if to meet the rising sun you will see that at the equator, that is at the widest part, a spot on the globe travels more miles per hour to the east than does one on the parallel of 40° North. A current travelling from the equator towards 40°N will, through inertia, carry with it some of the 'easting' momentum acquired at the equator, which will be more than the earth has at the 40th parallel. So the current has turned to the right, relative to the earth surface. So it is all over the world. Coriolis force tends to bend currents and force them here and there against the land, against ice barriers or thermal boundaries. The nett results of the systems is to give the semi-permanent currents, marked in any atlas, and semi-permanent up-wellings to be studied in oceanography. The currents are characteristically slow, about one mile a day or less. Winds which are sometimes claimed as causing the water currents, are subject to much the same forces: so there is no marked argument.

I would therefore name the Coriolis force the determinant of the life in the ocean. Without it I do not see any of that deep reservoir of rich water reaching the useful layers in the light of the sun.

General review
The first comment that I would make on this survey of aquatic environments is that if we compare Clarke's diagram of marine ecology Fig. 19 with my diagram for the farm—the fertility cycle Fig 11—we see that the appearance is similar. Next, let us note that, although each aquatic eco-system—pond, lake and so on to the ocean—seems to have its own determinant, they have many things in common. In all of them water movement and mixing are important to the living organisms, in fact, fundamentally so. In all of them mixing may be due to the wind, convection or to the movement of water bringing masses of two different densities close together, as when a cold stream runs into a warm pond or as when a river runs into the salty sea, or with the great movement induced by the Coriolis force and by the differences between tropical and arctic temperatures.

Then, in all systems there are important roles for aerobic and anaerobic bacteria. In all, the planktonic crustacea characteristically belong to the copepoda and cladocera and other groups similar in all systems. The diatoms and other microscopic plants which form their food are also characteristics of all waters.

In all the systems there are fish, which attract herons, cormorants and kingfishers, and finally, in all there are organic deposits on the underlying bed, although some deposits are not very thick, as in swift streams or in much disturbed oceans.

Hence all the characteristic interplay of biota and pH and redox are found widely spread.

There is, however, a great distinction in that the productive layer of the sea is many fathoms deep, down as far as light will penetrate and support photosynthesis, below which animals have to live by what drops down to them from above, and this they do.

When all is said, a difference remains in that whereas man has changed the face of the earth—in the sense of the land—out of all recognition, his effects on the watery part of the globe—what is called the hydrosphere—have as yet been small in relation to all the great volume of water. One can only hope that they will remain small and that even such effects as the reduction of whales and of fishes may be reversed by conscious control.

Pollution

No review of the water would be right without mention of river pollution. The problem is urgent in all industrial parts of the world. It is exacerbated by growth of population on the one hand and of pride in hygiene on the other. These cause heavy demands on water, both on the industrial and the domestic. Then the rivers are left short of water; whereas plenty of clean water is the only complete eliminator of pollution. The cure is carried out by bacteria and fungi acting on the contaminant, at high dilution with well aerated water. Contaminated liquid at high dilution is often difficult to handle, but a cure is sometimes provided by long rivers, e.g. Rhine–Maas, if they do their work properly, which is not always. I mentioned a few pages above that sometimes nutrient escapes from the Maas to pollute the sea and embarrass the herring fishery, but in my surveys of 1934–6, the shorter River Thames liberated more phosphate, and more often, as I recorded in two papers published in 1938. The River Trent, Dr. R. W. Butcher told me, cleaned itself twice during its long course.

Pollution by excess of nutrient salts, long thought by the engineers to be harmless products of sewage treatment, provides a special and difficult case, causing the whole water-mass to go green with weed or excessive algal growth, called bloom, the phenomenon being called eutrophism. In acute forms, it demands cultivation, or at least removal of protein, which can be in the form of those kinds of fish or shellfish that will feed on the algal blooms that the nutrients stimulate. It is pleasant to know that the shellfish at least can be made safe to eat if given sterile water for a short time. Canals or lagoons for the purpose could thus provide a new food supply derived from a new nuisance. But there must be a word of caution to guard against poisons, which animals sometimes accumulate.

For the usual case of general pollution, which is anaerobic, we should note that around the British Isles and probably round all rocky coasts in temperate zones, there is a vast quantity of turbulent sea-water. I doubt whether the cleansing bacteria would be overloaded if the water off all rocky, open coasts were used. Where people have thought it sufficient that the water merely be tidal, there are some filthy failures. Long Island Sound is the greatest I have ever seen myself. More instances need collecting, but favourable factors seem to be: temperate climates so that there is not a marked thermocline; rocky coasts, which are evidence of turbulent waters; and open situations readily refreshed by water from ocean or near ocean.

In some of the livestock areas of Britain, the drainage from farms makes a black comedy, to be read in the *Journal of the Farmers' Club* for April 1970. 'Slurry', which according to Chapter 4, should be the priceless essence of agriculture, plays the part of poison, in great quantity. In the discussion, the sea figures as the convenient dump, but it is truly not so handy as many landsmen assume. The cure, mentioned in the previous paragraph, may be far away.

The problem of pollution, therefore, seems to be only one of distribution and transport of the polluted liquid so as not to overload the areas of natural treatment. In the meanwhile, river pollution is dangerous, even menacing, and, some would say, disreputable; so is sea pollution in some sandy or enclosed localities.

Finally, there is the special case of radiation. It is important that many fish are safer to eat than would be expected because radiostrontium washes in and out of their bones, interchanging with the strontium of the seawater. Thus, like mussels with bacteria, when in clean water they cleanse themselves.

In general, conservation morality may come first upon the land, where forestry is all-important, and most hopeful for conservation. Let us therefore consider forestry.

Forestry 6

Properties and uses of timber
Like the sea, forests cover great areas of the earth, and like the sea again, to most of us they are remote and unfamiliar. Those who know them well are separate people, full of their own special lore of the forests, but the subject can be started by looking at the common wooden wheelbarrow, which many of us do know.

Barrows of steel bend and perish at the rivet holes, but a really good wooden barrow will last a lifetime. The rather cheap barrows made of various timber of not necessarily the best selected pieces can teach us more, as an introduction to forestry. One such barrow broke on me at a joint under the body. The break was in one of the long members ending on the handles. It broke because the grain was, as they say, 'short', whereas in vulnerable positions the grain should run along the length of the timber. If the grain is right, the timber will spring back into position when the load is eased. Such is the property of ash, one that makes it valuable for wagon or car-body-building. But oak of the right grain will do as well. The all-oak barrow may seem too heavy, but in many heavy duties its weight is not appreciable, compared with the load.

Even cheap barrows should have legs of oak, because resting on the ground, they are liable to be wet at the bottom, which perishes most kinds of timbers.

A third kind of wood to be seen in many barrows is elm, boards of which form the box or body at the front and sides. Elm stands hard chafing wear and resists splitting, because the grain is so often crossed, and as it were, woven. It does, however, warp and move with changes in the dampness of the air, which is another result of the cross grain. Breaking across is another. This timber is noted for lasting under ground, or under water, as in Roman water pipes, which could be made of elm trunks that had rotted down the middle, as elm often does. This tendency to rot if aerated makes it unsuitable for fencing posts, which should be of oak. Of elm, they would break off in a year or two, rotting between wind and water as the saying is.

So it is perceived that timbers have different properties and different uses. It is a remarkable achievement of carpenters and timber merchants in all countries where timber is used and in all those from which timber comes, that those men know which kind will be suitable for each job. The catalogue of this knowledge is to be found in most forestry books; I have for many years used C. O. Hanson's *Forestry for Woodmen* which I commend.

The problems in utilization naturally direct the forester's efforts towards growing timber to meet requirements, as far as he can foresee them, but he has to look a long way ahead. Oak is 120 years old before it begins to be usable; and is better at 200. The best for building and furnishing is grown in deep woodland on rich soil. When planted closely, it grows straighter and taller than it does in the open, giving longer pieces and of straighter grain (Fig. 28)—to come back to the wheelbarrow again.

Ash tends to grow tall and straight, although again it will give

28 Canopy makes proper timber. The canopy makes good timber because trees in the shade competing for top light make long trunks of straight grain and, unless there is also side light, there are no side branches to grow big knots and weaken the strength of the cut timber. Niches, here, may be noted, for further interest in Chapter 7 namely: (*a*) canopy; (*b*) trunk, bark; (*c*) side branches; (*d*) shrubs; (*e*) field layer; (*f*) leaf mould; and (*g*) roots.

that long springy strength only if grown fairly close, tree to tree, deep in the woods. In hedgerows it tends to branch out to make a shape like a bunch of flowers, such as the tree I am looking at as I write this. Such an ash will give straight and valuable timber only in short pieces, and as the grain is rather too open to give a fine finish or a good immoveable fit, it has not much use in width or when joined: we do not see doors made of it. But ash is good for gates, where the joints are rather coarser made and its resilience prevents its breaking.

Ship timber
The appearance of the English countryside has been greatly affected by the usefulness of oak trees in building wooden ships. Here it is the trees of the hedgerow that used to be of greatest value, surpassing the value of those trees grown deep in the forest.

It is said that Collingwood, an admiral colleague of Lord Nelson, in his retirement, used to plant acorns in hedges in his country walks, but not all the oaks in hedges have been planted. Rats, squirrels and others hide acorns there, which, safe from the plough, grow into those trees which make our land so beautiful. Few of those are cut down for use nowadays, but I imagine that a few growing near the ports where small wooden ships are still built find their traditional usefulness. For fastening ribs to deck-beams, brackets are needed called knees. Oak makes plenty of knee timbers—if you look at oak branches you will see many with angles in them and those natural brackets are much stronger than anything that could be cut out of straight wood because the grain in the oak branches curves round the angle. Also the ribs have to be more or less curved and many oak branches are so.

Virgin forest
Most of the timber that has been used in civilization has come from virgin forest, meaning something not planted by man, nor utilized in recent decades.

I first met virgin forest when I was looking for a beaver in Newfoundland. There were in fact no big trees; nevertheless, the area could be classed as 'virgin' because the ravisher was not man but beaver. Beavers live on saplings near watercourses and build dams to raise the water level, for safety of the lodge, where they breed and store timber food. Whilst in Newfoundland, I was anxious to see a beaver. My guide knew how to make the beaver come out of his lodge, simply by breaking down his dam and causing the water level to fall.

It must have taken us nearly an hour to damage the strong dam enough to set the water falling in his pond and it took the beaver about one minute to stop the water falling, when he came out. First we saw him dive near the dam and bring up some logs which my guide told me the beaver kept ready stuck in the mud for emergency dam repair. We saw him swim twice with logs in his mouth to the place where we had the small gap; then he used some small billet wood. He then dived down and came up with water-plant roots in his mouth which he took down, then finished blocking the gap by making a flurry of mud from the bottom of the pond. The water stopped falling completely. Thus, it was only a simple job for the beaver; whereas I seem to remember that as a boy I found it very difficult to stop water falling at a dam. All the time the beaver had been working he had taken no notice of us at all.

Thus, virgin forest to me suggests beaver undisturbed, and still more so the small birds. When the guide went away I sat down on a fallen log, and little tits, 'chickadees', finches and warblers came to the same log, paying no attention to me at all, and not in the least frightened. I realized then that this really was virgin forest.

Plant succession and climax

A place in Canada proper, where later we lived, had provided a quantity of fine timber about a hundred years before, the yellow pine of a climax vegetation. Climax is an important and useful word in ecology, and we might leave the Canadian yellow pine for a moment to consider the meaning of the word. It is used with another term 'plant succession', so that we have 'the climax of a plant succession'. The meaning of the phrase can be seen much nearer home than in Canada, in fact within three miles of Salford University. Fellow ecologists are surprised that we can learn anything about forestry within so few miles of the centre of the twin smokey cities, but the use and abuse of land has left many instructive pieces, where afforestation takes place without design or control, and all the more instructively for that.

In a good seed year for birch, which is not every year by any means, there are millions of seeds in the air and they will germinate on grass clumps such as form among moss. So it is that on one site near a railway line, where waste colliery shale and ashes have been dumped in fairly recent years, there are birch saplings. Further inward on the old colliery site there are older birches, one of which I judge to be as much as twenty years old. Crowding close to it, there is a sycamore sapling four feet high, which will grow into a tree when the birch falls, which it almost must do within the next twenty years. Still further inwards on the site, dating from before the colliery closed in 1896, there was also, in 1969, a great poplar, which had freshly fallen, one of half a dozen or so that may have been planted by the father of a man I met, who told me that his father, when invalided with silicosis, did some planting on that and other nearby tips. At the edge of the flourishing sycamore and alder wood, there is a 3 foot oak seedling. Thus a succession from birch is demonstrated, birch, sycamore, oak, but also with poplar as a pioneer, and alder as a staple. In the damper areas, the first trees to strike were evidently the sallows—pussy willows—or the osiers, which are also called pussy willows. Ecologists say that the oak is climax vegetation: when other trees become old and perhaps fungus-ridden and die, the oaks will still be young and healthy.

On soils over chalk, beech is the climax rather than oak. Mr. Valentine of the Bolton Parks Department pointed out to me some beeches commonly growing in clumps in fields round Bolton. He judges that no one planted them. He also notes the commonness of the field name 'croft', which meant a bleaching ground. He thinks bleaching powder left lime residues, and so there are beeches, rather than the usual oaks of the naturally acid soil over the millstone grit.

The word 'climax' is not to be interpreted too strictly. When the beeches die or fall they may be followed by yews, which some people think provided the bows for the Bowmans, of which there are many in the north-west of England. By the same argument, yews grew in the Bowlands, presumably on the limestone outcrops. However, without answering such speculative questions, it is convenient to let the term 'climax' include sub-climaxes, such as beeches here.

It has taken several paragraphs to explain what is meant when we say that in virgin forests the trees form a climax vegetation. Such were the yellow pines of eastern Canada, already mentioned, large slow-growing timber which was much used for building in England's industrial areas during the 19th century. They formed wonderful wood and a load from the dismantling of century-old buildings did good service about my farmlet.

Ladang

It is very doubtful whether what seems to be virgin forest in many hot countries is really so, and its trees climax vegetation. Doubt is thrown owing to the ages-old practice of farming it on the system called *ladang*, which many ignorant people scornfully regard as feckless. The local villages slowly migrate through the forest, burning a piece to keep them for a few years. They leave standing the dead stumps, amongst which they grow gourds and melons and other fruits for two or three years. Then they can grow maize there, but after another three or four years the soil is exhausted—very often it was only an inch or two deep—and so they move on and burn another piece, while the patch they have used recovers as forest.

A misguided British government had a grand scheme for tearing all the trees down, ploughing the land and sowing ground-nuts—'monkey nuts'. The natives had grown ground nuts by the ladang method: with the white man's method the forest came back so vigorously as to defeat all methods of weeding. So, the practice of ladang or 'burn and move on' proved to be the only one by which the land could be farmed at all. At least that must be the interim conclusion for the 'thin forest soil' as it is called.

Conservative lumbering

The next kind of forest to consider is what grows after felling or abandonment, that is, the naturally regenerating forest in the forest zones where trees grow like weeds. Familiar ones are mainly in the far north. In eastern Canada, in Nova Scotia and New Brunswick, one comes across abandoned farm areas. People went out, especially after the Napoleonic wars, to clear the forest and try to farm. Now, as you go through the fir trees, of all sizes from seedlings upwards, you stumble over old walls and find orchards that have gone wild. The impression is of nothing but fir trees, though there are also hardwoods, especially maples and birch. That sort of forest can remain profitable, because of conservative lumbering, in which the lumber men go through it at intervals of fifteen to twenty years taking the useful timber but never totally destroying the forest. Or acres of it go in succession for wood pulp, several acres for one edition of a Sunday newspaper, they say.

The greatest trouble is caused by fires: a forest changes over the years because of fires. The effect of a fire is devastating and extraordinary. Walking through an area which has been burnt, you may find yourself sometimes twenty or thirty feet above the ground, making your way over as it were a tangled heap of giant matchstalks, thrown down higgledy-piggledy as we used to throw down spillikins in a Victorian game. The fire evidently killed, and so felled, many more trees than it consumed. As to causes of fires, in very dry weather it is possible the friction caused by branches rubbing against each other could cause a fire; the sun's rays might easily be focused on a bottle left by people, and it is possible that bacterial action could produce heat in rotting vegetation just under the surface of the earth and that a strong, dry wind could fan this heat into flames. All these are possibilities but blame is often put on sparks from the railway.

In modern northern lumbering, there comes a time when it is not worthwhile going deeper into the virgin forest; it pays to use the naturally regenerated trees in nearer areas, more especially in places within reach of rivers, which are used to transport the timber. Although one can say that these forests live by grace of human will, they still have many of the attributes of the virgin forest; for example, it is very easy to get lost in them. I once felt a very helpless person in a Canadian forest, although I was not really far from the railway and towns.

I had felt so inferior myself that I was pleased later when my guide got lost, we were 'turned round' as they say, when they do not know in which direction they are going. However, he knew how to cope with the situation, which I did not. We went round in a rough circle and he marked trees as we went in case we should come back to the same place by mistake—you cannot walk straight in a forest because you cannot see anything but the trunks. But here we were circling deliberately. My guide was looking for a river or stream. When we finally found running water he looked at the direction in which it was running and then, from previous knowledge, knew in which direction we were facing. It was autumn and the temperature at night was dangerously below freezing point. I asked my guide what we would have done if he had not found his bearings. He explained that, to stay alive, the trick is to find an overhanging rock and make a little fire the length of your body alongside it so that it warms the rock, then having collected plenty of firewood you creep in between the fire and and rock. That way you can survive the night. The smaller animals keep warm by moving quickly when out of their burrows and soon die, he said, if held in a trap.

Trees for conservation
We may surmise that ecology of forests must resemble that of farms, at least somewhat, because fields go so easily to woodland. I find that the piece at Rothamsted that I mentioned in the first chapter is matched by a piece of former meadow in Rivington Park, near where I live. The army closed it during the Second World War, and trees took over. By 1969, some oak trees were twenty feet high.

It has long been recognized that farmer and forester have a common interest in soil quality, and the aim of the woodman has often been to improve the quality of his soil, with timber production of incidental importance. The natural process is as follows. The tree roots grow down through the soil to the rock or subsoil, and, as with herb roots, the carbonic acid made in respiration dissolves some of the rocky or other inert material, liberating mineral salts. The trees thus show what we have called already 'mineral efficiency'. The salts are used in the growth of the plant, and finally a surplus lodges in the leaves and so eventually falls to the ground.

The leaves on the ground when lightly mixed with soil form what is called a *mor*, and when more deeply mixed a *mull*, an example of which we encountered in Chapter 3 as the excreta of earthworms. It may not be generally realized that earthworms and their exploiters, moles, are common in woods, wherever the pH is of high enough number; and so are all the other groups of soil organisms in Chapters 2 and 3.

Thus, the vehicle of soil improvements under trees is the leaf-fall, building up as it does humus, year by year, and sufficient salts, always provided the leaves stay on the forest floor. It seems from my reading of C. O. Hanson's *Forestry for Woodmen*, which I have mentioned, that any kinds of trees until they are 20 to 30 years old, form sufficient protection for the floor, so that its covering does not blow away or wash away; but that after that initial decade or two the outcome varies with species and management, and it is possible for deterioration to set in. Restorative management is being investigated from the Merlewood station of the Nature Conservancy at Grange-over-Sands in North Lancashire, using plot trials in what appears to have been an over-exploited wood. I might mention that the Merlewood work in Roundsea Wood is over-run by roe deer. That is one of the forester's problems; it is also part of the ecology, and if I were a forester I would try managing grazing cattle to oust the roe deer and fill their niche.

Plantations

The English wood is usually deliberately planted; and I think this must make the northern lumbermen laugh heartily. There can be little prospect of English planting making a profit, when for oak we must wait 120 years for a return on the cost of planting, fencing and maintenance; for beech 90 years; for pines, larch, sycamore and ash 60–80 years; and even for firs, spruces and the like half a man's lifetime. Only poplars and willows will give the planter something to sell within a period that he might think reasonable, 15 or 25 years.

The truth is that people continue to plant trees for reasons other than financial gain, yet the art of forestry depends on attending to the financial aspects; and, after all, no one wishes to lose money by mistake.

The market for timber has a trick of varying unpredictably within the lifetime of the tree planted. I once helped my father to sell some wood off a few acres, and the merchant could not allow us any real money for the sycamores, there being no demand for that kind of wood. But by 1950 I was hearing of sycamore as making one of the highest prices of all English timber. Thus, I would think that present fashions and demands are hardly worth noting.

Instead, attention to qualities and specific suitabilities is perhaps safer, I have never known a time when ash, oak, or larch, would not sell.

In cross section, the age of timber can be read, because the vessels formed in spring are larger than the rest, carrying more sap. However, aberrations in the weather can cause difficulties in the subsequent reading.

After the suitability of the timber, or before it, there is the question of the site. When deciding what species to plant, owners give site and aspect a great deal of consideration. For example, ash is least frost-hardy, willows most so. Above 1000 feet a planter may choose larch, birch or Corsican pine. On north or east aspects, he could try ash, beech, larch, hornbeam or spruce. Sycamore is wind-hardy, but all the same it is included with oak and elm, as for south and west aspects. Much more of such advice is in the forestry books. Planting is such a long commitment that everything has to be thought of, if the planter is to reduce the risk of failure.

Volume of standing timber

In a wood many trees contain usable timber, the volume or 'cube' of which is estimated by a method which gives a figure $21\frac{1}{2}$ per cent less than the actual volume, to allow for loss in sawing, and another 10 per cent is often deducted for bark.

The rule uses 'timber height' in feet, that is, excluding root swelling and top, and 'quarter-girth' in inches half way up the timber height. The 'cube' is found by squaring the quarter girth and multiplying by the timber height, correcting for the inch measurement by dividing by 144.

If the quarter-girth cannot be measured directly half way up the height, it can be estimated by observing the rate of tapering. To measure the height the timberman can use an instrument called a hypsometer. This depends on 'similar triangles', one of the triangles being from himself to the top of the good timber and then down to the foot of good timber in the trunk, then from the foot of the tree to his own feet—with corrections for his height of eye. The instrument reproduces this triangle. It consists of a sighting tube to which is fixed a cross-bar, graduated in inches, which can slide up and down. On the sighting there are notches every inch in which can catch the line of a plumb bob which is suspended

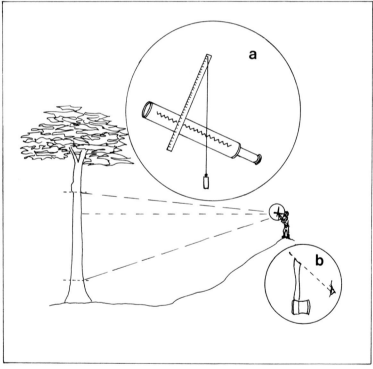

29 *Estimating timber height.* Two methods are shown: (*b*) was demonstrated to me by Mr. Pearson, a Salford Parks Officer. He suspended an axe from its handle, forming a plummet, and walked backwards until the base of the handle, which was cut at 45°, sighted on the top of the tree. If the tree were then felled, it would reach to just behind him, namely as far behind as his 'height of eye', as a simple drawing will show. Method (*a*) also uses the right angle triangle, through an instrument called a hypsometer. The sliding cross-arm is set at the correct graduation for the distance from the observer to the foot of the tree. The result is two similar triangles, as the sketch shows.

from the top of the cross-bar. If the cross-bar is set so that inches on it correspond to the number of feet from the foot of the tree to the observer, then the number of notches will give the height of the timber. In Fig. 29(a) I have assumed the observer standing above the level of the tree, when the hypsometer is used twice, and the height and depth added.

A still simpler device was shown to me by Mr. Pearson (Fig. 29(b) who looked after Kersal Moor, the ground where we were making the map, mentioned in Chapter 1 as the practical work of this course. He showed us that the handle of an axe is cut off at the bottom in a 45° angle, consequently if he holds the bottom of the axe handle between his finger and thumb, so that the head acts as a plumb bob, and walks backwards from a tree, sighting it until he has got the slope of the end of the handle aligned on the top of the tree, then he is standing nearly at the place to which the tree will reach when he fells it. Allowing for the height of the eye, it will fall his own height behind him. That is a very useful piece of practical information for men who are felling trees in confined spaces, and I had never seen it before. Obviously, you could turn that into a height measurement by determining the distance from your position to the tree.

There is a difficulty if you are unable to walk from your position to the tree, owing to a river or some other impassable barrier in between. You have then to use 45° triangles in a different way. If, from the position where you stand, you peg out a 3, 4, 5 triangle, based on the line from yourself to the tree, you will set out at right angles; if you then walk until your 45° instrument, whether it is the butt end of an axe handle or something that you have made for the purpose, sights the tree, the distance that you have walked will be the same as the distance from the tree.

So much for the volume of a single tree. A man who lives by buying standing timber, felling it and sawing it up, often has to value a wood. Then he must estimate the total volume of each kind of tree, involving numbers and average size. When little more than a boy, I watched an old and respected merchant do this, for a small mixed wood. He walked slowly through, and put a few figures on the back of an envelope. There and then he presented us with the answer of, I think, £430. This was about a fifth more than the best previous offer, but so good was this merchant's reputation that the offer was raised on hearing his estimate. By judgement as he walked he must have formed a mental picture of the average tree of oak, ash and sycamore, and have noticed how many he was passing in his measured pacing; and so obtained an estimate of the

30 Opportunity from fire. Cooper points out that these trees are orderly groups of pure stands of even age. Only the youngest group varies in age, because the last fire, which let in the light necessary for the germination of this species, consumed the combustible needles, consequently the seedlings for several years were untouched by fires, which otherwise might have been lit by lightning. The possibility of the wide importance of fire in ecology was fully discussed in the Wenner-Gren symposium in 1956, and indeed earlier in the U.N. symposium in 1951. *31 Order from fire.* A second picture used by Cooper. Fires occur at intervals of 3–10 years and so are mild, for lack of many accumulated needles for fuel. Fires kill the smaller saplings, only the strongest are left, so making the uniform groups of these Ponderoso pines, in the south-west of the United States. Foresters tend to be too nervous of fire and Cooper thinks that they should use it more boldly. In this drawing the fire scar at the foot of the one big tree has increased only slightly and the dead smaller saplings can be seen in the foreground. The light fires in the background must be distinguished from the devastating holocausts which he says are liable to occur from failure to use fire lightly and frequently. Figs. 30 and 31 are taken from *The Ecology of Fire*, by Charles F. Cooper, © 1961 by Scientific American Inc., all rights reserved

total number in the area of the wood. A less experienced man would need a more formal method. One could go a long way by using the pioneer statistical methods of Sir Francis Galton, which are useful and convenient in practice, in a way that the later highfalutin algebraical methods are not. Thus I would commend finding the median oak, which is one with an equal number larger and smaller than itself and measuring it up to represent the average oak and the same for the ash and the sycamores. Thus I would have to measure three trees only, and very often the answer would be quite sufficiently accurate for the task, because of the uncertainties in sawing out. However, there are Forestry Commission methods, which anyone interested may find from a handbook.

All this may seem rather remote from the life we ordinarily lead, mainly in towns or cities, and anyone might well wonder why a general ecologist should be invited to be familiar with the volume of standing timber. The answer is this: everything alive in the wood is affected to a greater or lesser degree by what the forester does there; and everything the forester does in the plantation aims at a greater volume of timber—it is his determinant, and therefore the key to the plantation. It is one of the instances of determinants in ecology. The determinant of a forest is the volume of standing timber. Without that aim, the wood would be a kind of jungle, or at the least a jumble.

It may be worth mentioning some of the actions in managing an English plantation. Some English wood grew up by regeneration from stools, or from seeds from mother trees which had been left standing for the purpose when the forest was felled, if clean-felled, that is. Or, if the forester simply takes out the trees as they become ready, without disturbing the forest he may obtain the best conservation effect, because the forest soil is not disturbed nor do bare areas invite weeds to grow up. However, an objection is raised that the quality of the plantation gradually diminishes, by loss of the fastest growing specimens.

In order to obtain the greatest possible height clear of knots, the forester plants his young trees closer than necessary, often at 4 feet centres, and will, when the time comes, thin them out so that they shall make big timber, but in the meanwhile his trees should have committed themselves to growing tall and straight uniformly, with grain accordingly. He may also have 'brashed' the trees, as it is called, that is, cut off the side branches.

Foresters sometimes provide nesting boxes for tits, which feed on insects, some of which are harmful to trees.

There is much more in this extremely interesting subject than I have given here. My aim has been to give an added interest to country walks and added awareness in planning the countryside, and added sympathy for the people who have taken the trouble to plant for the benefit of those who come after them. Biologists, however, and full citizens, should want to know more about what grows in the woodland and how the living creatures order their lives, in fact to turn, as in the next lecture, to more general woodland ecology.

7 Woodland

A book

In each chapter so far, I have found myself making new major points, that is, points that are not to be found elsewhere. The same will be true of most future chapters but in this one the main ground has already been well covered, my task being one of adaptation to the special study of determinants with which the present book deals. There exists already a fine small book describing the ecology of woodland intimately: E. G. Neal's *Woodland Ecology*.

Habitats

Neal points out first that the wood contains various habitats, some very different from each other. He instances the little pools that are sometimes to be found in the boles of trees and which contain pond life in miniature. He does, however, advise against making too many habitats in the study, otherwise you end up by putting the same animals in so many of them whereas the main justification for the conception of a habitat is that it contains a distinct collection of animals and plants.

In the preceding chapter, we walked through a wood with a forester, and learned of his problems, and we are now going to walk through with a naturalist and learn something about a wood as he sees it.

On entering a wood we automatically stop to look up in awe, very much as we do on entering a large church or cathedral and for the same reason, namely, the suspension in the air of solid objects high above us on an intricate and admirable tracery of members. In the wood this is called the canopy and is like an awning of leaves and fine twigs carried on branches, which are big and strong at the base, but become smaller and smaller as the canopy is approached. The canopy is a favourite resort of certain animals in the warmer countries. It is where the butterflies and monkeys are, and it is thought that this is the habitat where primates acquired their sense of colour, which is not well marked in other groups of mammals.

In English woodlands the canopy is hunted by blue tits, which in summer seek the numerous aphids on the underside of the leaves, and scale insects and other inhabitants of the high twigs and branches. It is the home of magpies who make their nests so high that it is a real test of a boy's courage to go so far up and on to such slender twigs in order to collect an egg.

All woods tend to form a canopy, even when the species of trees are mixed. A tree that is not growing so fast as the others will find itself in less light and will put all the more energy into growing up to a great height to obtain the brilliant light available in the canopy. In this way very thin rods of trees are sometimes found which are thinned out by foresters because they will never make great trees, but they are quite useful in country work such as hedging, and wherever a thin and whippy piece of wood is wanted. The result of competition to be in the canopy's light is that the trees reach a moderately even level and maintain it. Thus they 'mould', or smooth out, the contours of the land beneath.

The canopy in Fig. 28 (in Chapter 6) is what is called close canopy, where the twigs and the branches of adjacent trees touch or intertwine with each other. This is characteristic of, for example, the beech which, being a tree that is not stopped by shade, seems insensitive to some of its leaves and branches being obscured by others. And, as its leaves are round and have no serrations in them, they do form a very close canopy, comparatively impervious to light. Oak trees are not always grown in close canopy, and, even when they are, they let more light through because the leaves have wrinkled edges. Ash trees let still more light through because there are spaces between the individual leaflets of the pinnate leaves.

An interesting case is provided by evergreen conifers, the pine, spruce or fir. When these are grown commercially they are planted fairly closely and kept that way all their lives, so there is a close canopy which, being evergreen, lasts all through the year and allows very little to grown underneath. When the evergreen woods are growing naturally, however, as in the acid heaths in the south country or in Scotland, there are usually a good many spaces and there may even be birch trees with them. So too, heather and bilberries will grow below them. There is even some bracken.

As well as its density, the duration of shade is of great importance to the plants below. Beech has a long shade season. It begins to put out its leaves early, towards the end of April, and keeps them until fully the middle of October, and perhaps for longer into the winter. The same applies to oak, which, although it does not put out its leaves until May, does not lose them finally until November, but we have seen that its canopy is not quite such a serious shade as that of the beech. The ash, which we noted as open-leaved, is also of short duration, opening after the oak. In spite of the common saying, it nearly always opens after the oak, and it loses its leaves at the beginning of October, so that ash woods are not severe on other vegetation.

The trunks of the trees form a separate habitat and where there is a deep bark this itself is remarkable for sustaining certain characteristic fauna and flora. Where it is a little bit loose, or where branches are dying, fungi will get in and make it looser and beetles and other insects will then get under the bark and caterpillars of various kinds will start to eat the wood. The woodpeckers with their strong beaks will bore down and get insects from out of the wood itself or from beneath the bark, and even if the bark is quite untouched it does form a habitat for certain plants and animals. On the north side it grows the *Pleurococcus* which has been mentioned in previous chapters, and Neal tells us—I confess that I did not know it till I read his book—that slugs at night ascend the trees to eat *Pleurococcus*. Lichens also grown on the surface of the bark, and the tree creeper explores the cracks for small insect food.

Where the canopy is sufficiently open, one finds the habitat known as the shrub layer, which also includes any wood being grown for coppice, such as hazel wood. There you will find willow wrens at work on the aphids, as do the great-tits. Blackbirds nest in the shrub layer and so do wrens and other birds that tend to hunt more in the field layer below. Squirrels of course are getting nuts there, although they do get a large amount of food from higher up in the trees.

Provided that the canopy is not too close, there is next the field layer, which includes all the characteristic woodland plants. The grasses which are found in the fields are represented here by other species of the same genus, such as the wood soft grasses and a rather slender meadow grass. There are a number of characteristic plants such as bluebells, and endymion which survive by stores of food in their roots, because they must undergo a dormant season in the summer when the canopy gets too thick to allow them to continue full production. Here you will find the wood mouse and the field mouse which can find a living on the vegetation and on things that drop from the layers above.

Finally, extending deep into the wood, there is the leaf litter and leaf mould and soil layer, taking them all together. Here the badger and fox have their burrows which are called by special names, and here the nourishing processes corresponding to those that we have studied on the farm go on. From Chapter 6, we may recall that forest soils, when they are dry and leafy, are decribed as belonging to the group *mor*, whereas when the humus has worked down deep into the soil they are known as mull. There are, however, some intermediate ones; so all soils cannot be distinguished as belonging to one class or the other, useful as the names are for the extreme conditions.

Russell, in *Soil Conditions and Plant Growth* generalizes to this extent: that the mors are worked down by mites and fungi; the intermediate soils by millipedes and centipedes, woodlice and springtails; and the mulls are definitely the work of earthworms and provide their food. The final form of each of those categories depends, he thinks, on the fauna that are working in them; all in their different ways breaking down the vegetable remains to form humus.

There are other habitats that could be considered, for example I am inclined to notice myself the difference between the canopy and its fauna, and that of the side branches. Near the edge of the wood and in gaps there are flourishing side branches above the shrub layer, but the fact that Neal does not list it as a different habitat suggests that he did not find the differences in biota worth separate attention.

Niches

Having dealt with some of the habitats that we have distinguished in the wood, it is of considerable interest to note that while the habitats appear reasonably uniform, each animal or plant lives in its own world, making habitats by the thousand. These are called niches.

Let us consider the blue tit. Its niche is that of feeding on the undersides of leaves in the topmost twigs of the canopy and on the outer twigs of some other branches. Light in weight and extremely agile, it is well adapted for picking off the aphis and other insects that are under the leaves. In Neal's Thurlbear Wood, even in winter, it seems to have been able to find enough aphis eggs on leafless twigs to survive, so that the feeding part of its niche was not apparently a constraint on its numbers. The constraint came from the fact that it does not make a self-supporting nest but keeps its numerous eggs and its agile and clamouring fledglings down a safe hole. The number of blue tits in the wood seemed to depend on the number of available nesting holes. This reminds us of a practice mentioned in Chapter 6, namely that foresters put nesting boxes in the trees so that they may increase the number of blue tits up to the limit and reduce the insects which damage the growing parts of the trees.

The niche of the great tits is rather similar to that of the blue tit, but being a heavier bird and less agile, it is at a lower level where the twigs are a little stronger. It, too, nests in holes which tend to be rather bigger than the holes of the blue tit. The so-called leaf-warblers give a similar stratification. They too live on aphids and other insects. The wood-warbler or wood wren is more fully arboreal than any of the others—it works at a higher level in the trees. The chiff-chaff which is the smallest of the three, works nearby at the same level as the wood wren, on side branches for example. Then we come to the willow warbler or willow wren. We see more of them than the wood wren because the willow wrens work the shrub layer, and the field layer to some extent, and usually nest in the field layer in patches of grass but sometimes in the shrubs or even higher.

This sort of fine distinction between niches is characteristic of species and is one of the factors taken into account in defining a species.

I have only in my life found or defined one new species of animal and that was a fish. I remember that I felt very much educated in zoology generally by the reaction of the British Museum authorities, Regan and Norman, when I told them that one of the forms I was distinguishing seemed to live more inshore and in shallower water than the other. The depth difference was only a foot or two, but it was clear. They at once said that a habitat difference of that kind clinched the matter of distinguishing the species, indicated by such anatomical differences as I had pointed out. There were, incidentally, two different native names, showing that the local inhabitants had already distinguished the species correctly, which the scientists at first had not done. It looks as if we can take it as a principle in ecology and in systematics that each species occupies a slightly different niche from its near relatives.

Regan's well-known and joking definition of a species, that it is what a competent systematist says is a species, is true as far as it goes. Another definition is that when some characters are measured statistically, there is either no overlap between two populations, or else there is a distinction in the averages of the two which meets some statistical criterion of confidence. That also, I am sure, is perfectly sound, but what both amount to is a skilful way of distinguishing species which occupy separate niches. Because they occupy separate niches they do not usually interbreed and this causes an isolation of the two populations and allows the genetic constitution to alter in one or both of them to such an extent that the populations can no longer interbreed successfully, that is, they cannot have fertile offspring. The skilled eye of the systematist or the careful measurements of the statistician do no more than indicate that this effect has taken place or is on its way. The fundamental cause of the separation is the occupation of separate niches, because some enterprising individuals or groups have found some comparatively little exploited source of food or shelter.

Niches need not necessarily be distinguished in place. They can vary with time. The studies of Thurlbear Wood include a graphic account of the oncoming of night, with first the crepuscular fauna, the badgers and bats, with the first owls, and then moths active and the nightjar following them. In the same habitat, there may be a diurnal niche and a nocturnal niche occupied by different species. The top bird of the day was the sparrowhawk, the top bird of the night was the tawny owl.

Very possibly the fox, badger and rabbit are only crepuscular or nocturnal because of man's hostility, their niche being one of safety. There is nothing in their staple foods that would necessarily make them nocturnal, compared for example with the bat, which has to chase the insects that are about in the cool of the evening, or with the nightjar which feeds on moths, at night, when they are flying safe from most birds.

Neal tells us, by the way, that owl pellets often contain bats' remains. Utilization, even by owls, in this way, is an example of the principle of efficiency arising from the niches. Neal is able to testify that he could not find a potential source of food in the wood that was not utilized by some species or another.

Pyramids

In our study of aquatic ecology in Chapter 5 we showed that in ponds the fundamental production is of weeds or of phytoplankton and that the various food connections lead up to the pike. In the wood, the corresponding basic production is in the leaves, whether in the trees, the shrub layer or the field layer. These form the food of aphids and of caterpillars and the aphids in their turn are food for ladybird and hover fly larvae and even for a kind of bug that goes over them, piercing their bodies in quick succession and sucking out their juice. Aphids are also eaten voraciously by the birds. Even a sparrow in the aphids season will feed its young on aphids, although for the rest of the year it is strictly vegetarian. The birds themselves—the willow-wrens and tits—formed the prey of the sparrowhawk. (Unfortunately, the sparrowhawk had become a rare bird in England by the 1960s.)

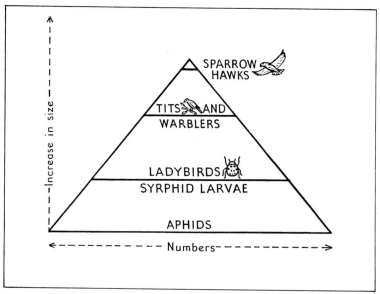

32 *Elton's pyramid.* Neal's version of the well-known Elton pyramid which shows that as one proceeds towards larger animals and stronger predation, the numbers are reduced. In 1970 Dr. Neal told me in a letter that the fauna of Thurlbear Wood was not greatly altered, except that buzzards were more often seen. Figs. 32 and 33 from *Woodland Ecology* (1958) by kind permission of Messrs Heinemann Educational Books.

Charles Elton pointed out that the numbers of animals at each layer form a pyramid (Fig. 32). In Thurlbear Wood, which covers rather more than thirty acres, there must be millions upon millions of aphids, thousands of ladybirds and hover fly, hundreds of small birds, and only one pair of sparrowhawks.

The aphids form what Elton has called the 'key industry' here, just as in the fallen leaf area the mites and springtails are innumerable and start the utilization of waste products, namely the fallen leaves. In the area of mor, bacteria and fungi have to work on the fallen leaves before mites and springtails can tackle them, but in the mull part of the humus layer, earthworms can work on the leaves directly.

The smallness of the number of animals at the apex of the pyramid may be illustrated by foxes and badgers, which at Thurlbear could not rely entirely on the resources of the wood but had also to hunt the neighbouring country.

Sometimes a pyramid can be quite a short one. Rabbits eat the leaves of the field layer, and there are some scores of rabbits in thirty acres no doubt, but the rabbits form the food for less than one pair of foxes. Sometimes the pyramid gets away from straight geometry, as when caterpillars are particularly successful in defoliating the spring leaves of oak trees and the trees correct for this mischance by putting out a second crop of leaves in the summer.

Consideration of these pyramids leads to several interesting questions such as what would happen to the leaves if it were not for the predators above the aphids. There is the question of healthiness of the prey population itself, unless thinned out by a predator that is selective upon the weaker members. That is a large and very interesting matter, which we must carry forward to a later chapter. There is also the question of what happens to the top members in the pyramid, and rather complex relationships can arise from consideration of parasites. There are also other questions, for instance, what happens to squirrels? The marten, which feeds almost entirely on squirrels, is a rare animal, and I suppose that if the squirrels are not thinned out they may become old and unhealthy and perhaps get caught on the ground by a fox.

Cycles
In woodland, as in other ecology, the living animals depend for their maintenance on those among them which specialize in the decay of waste products and of corpses, if any.

Neal distinguished the dead matter as dead plants, dead animals and dung and although there is some overlapping between them, those three categories do form useful entries in his diagram which I reproduce (Fig. 33). This might be called half of the cycle of decay. I show a one-way route and in reality it comes back again to the production of live plants and animals via the activity of bacteria and the liberation of nutrient salts, which two activities go on at all stages. Reality is much more detailed and complex than can be shown in a diagram like this, but it does bring out one philosophical fact. There is not one of the conspicuous and no doubt proud animals of the upper echelon, namely the owls, hawks, bats, foxes and badgers, that is not nourished in some degree by the activities of lowly beasts such as earthworms, woodlice and dung beetles and the other agents of decay. In particular the fungi, which are of the very first importance, do not lend themselves to inclusion in the diagram. Even the larger fungi may be eaten by a number of different animals; slugs can eat some of the larger ones that are poisonous to mammals and fly larvae and beetle larvae will also feed on them, even tunnelling into the bracket fungi that project from the trunks. Badgers will dig for truffles. The beetles

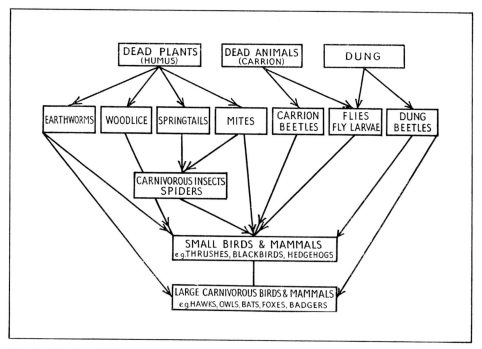

33 *Food chains and decay connections* (saprophytes) from Neal's Chapter 8.

and maggots feeding on fungi form the food of birds.

It is important that bracket fungi run their mycelium between the wood and the bark of dead branches and so loosen bark and let beetles and other insects live below it and form food for woodpeckers and other insect-eating birds, quite apart from the fact that the mycelium itself is edible for many insects. Among the agents of decay the little dung flies must be of great importance if only because of their large numbers.

On one occasion, A. C. Hardy came on a voyage with me because he wanted to discover what insects were in the air over the sea, moving in this way from country to country 'aerial plankton', he said. We flew tow-nets in the air from kites and when Hardy later identified the examples which we had caught, he found that they consisted almost entirely of dung flies, and this right out in the North Sea, two hundred miles from land. In the woods the dung flies provide food for spiders and dragonflies, which in turn are eaten by small birds, moles and shrews, while the earthworms which also feed on dung are the staple food of shrews, moles and hedgehogs.

Air pollution

Some of my correspondents elsewhere were sceptical about studying woodland ecology in Salford, and they had this much in their favour: that woodland shows clearly the effects of air pollution. The bark of the trees is of a darker colour than it ought to be—somewhat towards black. The particular places where air pollution shows most are those which are very rarely washed by rain, for example the wool of sheep, the bark of trees and the protected side of buildings, such as the eastern walls of Pendleton Church. The occurence of lichens on trees is an effective index of clean air and probably better than measuring soot and dust deposits, because it takes into account all sorts of factors. Another index, by the way, is the occurrence of house martins. I cannot explain that, but am interested in my colleague, M. J. Parr's observation of a pair of this species in Pendleton, Salford in 1969, something he had not seen before, since his coming there in 1955.

Tree trunks and sheep's wool may stay for a long time without

34 Air pollution. This is an oak tree at Rivington in Lancashire, showing, as many of them do in the district, the death of the crown, which is usually ascribed to smoke and sulphur in the air. This tree also suffers from erosion near the roots, but appears likely to survive for many years to come, a monument to Nature's strength against human fecklessness (from a sketch by Gillian Hunt).

being washed and doubtless that is why they show the dirt. At first glance some of the south Lancashire landscape may look clean, but usually it is not quite so.

The most noticeable effect of air pollution in the woodland is the 'stag-headedness', as it is called, of oaks. In a structurally charming clough on the edge of the moors, just behind my house, there are many oak trees, but none finishing with live branches pointing to the sky; on every one the crown of the tree is dead (Fig. 34). I ascribe this to the fumes from a nearby drainpipe works where four or five old-fashioned kilns smoke away, and the products drift across into the clough. The sheep nearby are all surprisingly black. I thought, at first, that my neighbour had bred them for fancy.

Naturally, oak trees that look like that are not healthy, and in fact the wood is brittle. When I cut a bough for a children's playground I was making, the bough broke right off, but it looked healthy enough when I carried it into Bolton. Oak should have been too strong for children to break, but that one was soon smashed. On the landscape, the stag's head to the oak tree breaks off and then the tree recovers so that after a decade or two we have bottleshaped oak trees, of which there are some interesting specimens close to my house in Rivington and further up the road. They certainly look peculiar, but they do not look unhealthy otherwise. It seems that it is only the branches up in the sky that cannot survive. We are four miles down wind from Wigan.

To see woods with their natural charm, it is necessary to go up-wind to Cheshire or far enough down-wind to escape the industrial fumes. The country changes into the beautiful as soon as one gets north of Preston.

The question does arise of whether there should be planning for further expansion of population in the air-polluted areas. The only complete cure is to go over completely to gas, washed free from sulphur, for all combustion, and although this would in many ways be inconvenient, it may well be that future generations will insist on doing so for the sake of having clean industrial areas such as the quite attractive ones in the Rhineland where I saw Kropotkin's idea of *Fields, Factories and Workshops* borne out. Whether that would be possible with the high rates of population per acre which exist in industrial Britain is a matter of some doubt, and I am afraid that a great many of the provisions for heavy density populations will be found in the next generation to be redundant, because the population will drift away. Even if we get the atmosphere clean and people's litter habits rather better, it is very difficult to get the same pleasure out of recreative facilities, including the countryside, if they are over-crowded.

General remarks
To return to woodland in general, the essence of it is surely the over-mantling canopy. Variations in the canopy control the nature of the lower layers of vegetation, and the insect, bird and mammal life in them. The determinant of a wood is its canopy. If a student sees the problem of the determinants another way, and reasonably points out that the subsoil governs the character of the wood, as can be seen on limestone outcrops in the gritstone parts of the nearby Pennine foothills, he gains full marks from me, and he may also cite climate, aspect and elevation. Reality is, indeed, even wider than the argumentative student sees it, on which point I would like to quote the final words of Neal's book. After mentioning the intricate links with other surrounding communities, of which I have just given one example, he is 'led to the conception that all life is really a unity—a magnificent complex of interrelated parts, the full beauty of which is beyond all powers of comprehension'.

I am sure that I must agree that the full beauty of the unity of ecology is beyond all powers of comprehension, in the sense of a comprehensive account being beyond our powers, but I do not think that it is beyond the powers of, shall we say, our apprehension, if we will accept and study determinants and as many of the bridging processes as we can understand. Migration certainly provides one of these links between eco-systems, one which we are almost compelled to study, and I propose to do so in the next chapter.

8 Migration of birds and fish

A part of ecology
When I have lectured to students on this subject, I have on two occasions opened by mentioning Philip Oyler's view. Philip Oyler, whom we met in Chapter 4, is no dreamer and is almost the furthest from being a fool you could conceive, whether as mechanic, builder or farmer. This superbly practical man holds that there is no mystery about animal behaviour, including migration, because God has provided the animals with whatever knowledge or skill that they need. It may be so.

My aim, however, must be to expose the facts as they are known, or can reasonably be surmised, and to give such scientific explanations as are available, which do in fact cover more of the field than is commonly supposed.

To establish that migration is an important part of ecology, we need go no further back than the preceding chapter, where we found the arrival of the leaf-searching warblers bringing to a halt the swarming and destructive populations of aphids. There are always some connections between eco-systems. If large, they should be taken into account in the two eco-systems in which they play their part.

In this chapter, birds and fish provide more than enough material on migration. A good authority on bird migration is Jean Dorst, and I use his fine book, *The Migration of Birds*.

Tern and plover
The Arctic Tern breeds within the Arctic Circle, that is, in Alaska, Canada, Greenland, Iceland, Norway, Siberia and Nova Zembla. It spends about half the year in the Arctic and the other half ten thousand miles away in the Antarctic, where, during the southern summer, there are rich supplies of plankton in the waters bordering the ice field (Fig. 35). In the northern spring the terns return to the Arctic to breed, where now it is the northern plankton that is swarming. This is surely the supreme example of making the best of both half-worlds. It seems almost as though it is essential that good use should be made of everything, and that Nature cannot

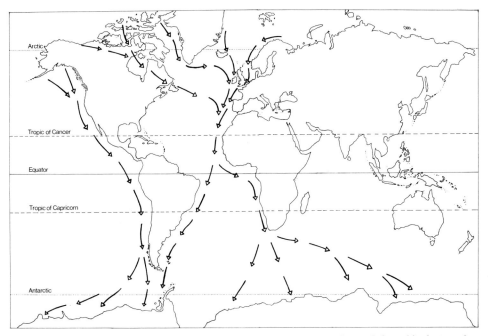

*35 **The migration of Arctic terns.** The Arctic terns avoid the long nights and the cold of winter in the high latitudes of both the north and the south. They thus crop their summer plankton twice in a twelve-month period. To do so they perform prodigious feats of flight which must also require great determination, in this graceful slender bird.*

bear that there should be quantities of plankton, anywhere, going waste, and so after its tremendous feat, the tern feeds lavishly in both the Arctic and the Antarctic, and avoids the insupportable rigours of winter in high latitudes, both north and south. The prize is rich indeed; but the price is ten thousand miles of flying twice a year.

The energy is not much less, and the apparent fittingness even more, for the American Golden Plover, which migrates every year between the Arctic and the Argentine pampas. The adults go south round by the coast but going home take a short-cut overland; whereas the young retrace their trek along the coast. The significance of the difference is not explained.

Landmarks
European migrant birds, the swallow, willow wren, chiff-chaff, garden warbler, blackcap, whitethroat and all those others which come every spring and summer and eat the insects, go for the

winter to various parts of Africa, where they can find insects after those in Europe have died down.

For many decades, men have wondered how the birds find their way on their long journey. On this question, it has long been noticed that birds become lost in fog. It was in foggy weather that small birds settled on our research ship in the North Sea, 'taking a passage with us' we said, but as soon as we could see the coast again they left. That is to say, the birds find their way by sight. For many decades, light-house keepers have reported the same and have come to the same conclusion. There is no point in looking further or in inventing other senses. There seems no end to the hypotheses that have been proposed, but until one of them explains why birds are lost in fog they are less valuable than the theory that birds find their way by seeing.

Given then that birds use sight, the next question is to ask what guide-posts they can use. Here we need not start from ignorance, because in the training of homing pigeons, from time immemorial, birds have graduated from short local flights to longer ones. This well established practice indicates that racing pigeons can recognize landmarks near home, and farther afield use other visible signs such as the shape of the land, edges of forest and the coastline. On being released, pigeons fly round gaining height before setting a course, evidently looking round.

Quite confidently, then, we can surmise the guidance for other birds such as the returning swallows. We can fly in imagination with them and see successive pictures as they approach. First the Guinea Coast, for they sometimes fly at an altitude of ten thousand feet, a fact which in itself makes orientation feasible, without any 'sense of direction'. Then comes the Moroccan Atlantic sea-board, the Straits of Gibraltar, the Bay of Biscay, the Severn estuary; all of which have recognizable shapes. Then, flying lower, they could find the estuary of the Ribble, the village of Longton, Mr. Mellor's poultry field, the third hen cabin in the eastern row, and finally the remains of the old nest under the roof. The practice of ringing swallows has shown that something of this kind must happen. Country people, whether they thought of Philip Oyler's divine provision, or had sufficient sense not to under-rate the birds' brains, have long surmised something like that itinerary, from their observation of the confident demeanour of the birds on arrival in spring, very tired but recovering to defend their claims by superb aerobatics.

Skymarks, migration fidgets

There seems then no great difficulty about the powers of birds to follow the pattern in daylight. However, we have also to account for the fact of birds making flights across deserts and seas where there are few features to guide them. Even if we assume none, the birds will not necessarily be lost.

In recent decades research workers have found that if you capture some warblers, for instance, in the early summer, they will settle down quietly in an aviary until the time comes for the free warblers to migrate. Then the captive ones will become afflicted by what is called 'migration fidgets', that is, they are on the move all the time, flying and hopping in the aviary.

A miller once told me that eels act in the same way. When the eels are running in the river there is activity in the live box, in which the miller is keeping eels to send to market. Mature eels run down the river in the autumn, especially at the dark of the moon and on stormy nights. The eels in the live box may not be able to see whether there is a moon or not, nevertheless they get migration fidgets, and when the miller hears the eels splashing about in the live box he knows that his trap in the river will fill up.

Sun as guide

Returning to birds, let us look now at Kramer's experiments with starlings. At the migration season, he used a cage in a round pavilion which had shutters in the otherwise dark roof. The following tests he performed repeatedly. (1) With the shutters open, he observed that in its fidget a starling tended to move to the part of the cage which was in the migration direction. (2) Kramer then altered the shutters and used mirrors until the sun, or its light in a slightly cloudy sky, appeared to be in an artificial position. The starling altered the direction of the fidgets accordingly. (3) One starling was tested at a time, observed through a transparent floor, and a mark was made on a diagram every few seconds to record the changing position of the bird. He noted that the starling was not confused by the time of day and worked correctly on the morning or afternoon positions of the sun. (4) In densely cloudy conditions the starling showed no evidence of any particular direction in the fidgets, but even the glimmer of diffused sunlight put it right again. Thus, without doubt, the starling was aware of the time of day and could guide itself by the sun accordingly. If Kramer's starling could do this, surely all birds could in some degree and keeping direction over trackless wastes is no longer a mystery.

R. Harden Jones in *The Migration of Fish* admits some degree of such power in fish on the basis of published work on coral fish, after finding something to criticize in Hasler's earlier work on white bass, which pointed to that conclusion. Dorst notes that ants and bees also possess this ability to correct for the sun's daytime movement, so it would be surprising if fish and birds had not also the power.

Stars

Ordinarily, most birds are unwilling to fly after dark, or even to move at all. Every farmer knows that at night roosting hens can easily be picked up from their perches. Yet some normally diurnal birds migrate by night as well as by day, and still find their way which is surprising in itself. It is even more surprising that the birds of the day should still be able to find their way in the dark. Sauer experimented with three species of warbler in a similar way to the experiments on starlings conducted by Kramer. (1) His warblers stayed awake in the migration season and fidgeted in the right direction in the cage, when they could see the starlit sky. (2) If the stars were obscured by cloud there was no direction in the fidgets. (3) Sauer then transferred the cage so that it was under the artificial sky of a planetarium, the lesser whitethroats again headed south-east which is the right direction for their migration, but (4) when the dome was altered to show the sub-tropical night sky the whitethroats altered the fidgets to head south, which is the change they normally make on reaching the latitudes where those stars are visible. Finally (5) he carried the birds to the latitude of their normal destination and he reported that at the sight of the southern stars the birds settled down and lost their fidgets although they had made no migration but had been carried.

These are the kind of experimental results obtained but many field observations support the conclusions in a more massive way. To sum up: birds travel by sight; by landscape recognition; by the sun or the stars; and thus vast movements of birds take advantage of seasonal supplies of food in many parts of the world. The determinant of bird migration is seasonal variation in food supplies—breeding migration being interpreted as homing only to the supposed ancestral home of the species.

Spawning and feeding

If we knew more of the migrations of fish, they might rival the birds in interest. The eel is one that we do know something about.

We think of eels as inhabitants of dark and often muddy waters, lurking hidden in the bed of stream or lake, and even in clear waters they have a strong instinct to hide among stones or any other obstructions. For several years of an eel's life it stays thus, hidden for the most part, although it is willing to travel at night and it may be caught by various cunning baskets or the like which utilize its liking for being protected. In general, however, the eel is regarded as local to the pond or stream where it is known to be. But in the autumn, in the dark and stormy weather, especially on moonless nights, the eels start to migrate downstream and can be taken in great numbers in traps built of gratings at water-mills. They will now have changed their colour from brown and yellow to brown and silver, their eyes grown bigger, their bodies more slimy—in fact you might doubt whether they were the same species as the eel you know. It has been with you for about eight years of its life, for males, or fifteen for females. But that simple statement can, it seems, be put in a more complex, and surprising way, thus: some individual eels mature and migrate young; these will be males; others stay in fresh water and become females, changing and migrating several years later. This and more strange facts will be found in Leon Bertin's book *Eels*.

The descending eels are clearly prepared for spawning, but it is reported that none is quite ripe. As they go down the river you may say a final goodbye to them, because they are heading for the ocean and will never be seen again. This is somewhat dramatic; they have never been known to spawn in fresh water and they breed but once far away in the sea and die.

In the spring of the year there is quite a different event, namely thousands and thousands of little elvers, called at first glass-eels, entering estuaries, such as that of the River Severn. Many are caught so as to make an agreeably edible dish of elvers, or to be sent alive in order to stock ponds and other fresh waters of Europe.

The gap in our knowledge between the silver eel and the elvers has had to be revealed by researches on the high seas, notably by the expensive and time-consuming expeditions led by Johannes Schmidt of Copenhagen in the research vessel, *Dana*, financed by the Carlsburg Foundation.

The ascending young eel has passed nearly three years in the sea as a larval form known as the leptocephalus. This is planktonic and like a willow leaf in size and shape, with little power of swimming. A series mounted in an exhibition leaves no doubt that leptocephali become glass-eels.

When Schmidt started, leptocephali were already known from

collections of plankton, particularly to workers from the biological station at Naples. In the Mediterranean, the larger leptocephali were taken near Naples and farther towards the east; westwards towards Gibraltar they were smaller, indicating that they entered from the Atlantic and grew in the Mediterranean.

When similar observations and reasoning were carried out into the Atlantic, it seemed that the spawning of all the eels of Europe and of Atlantic America is in the Sargasso Sea area, as remote as it could well be from the rivers in which they grew up.

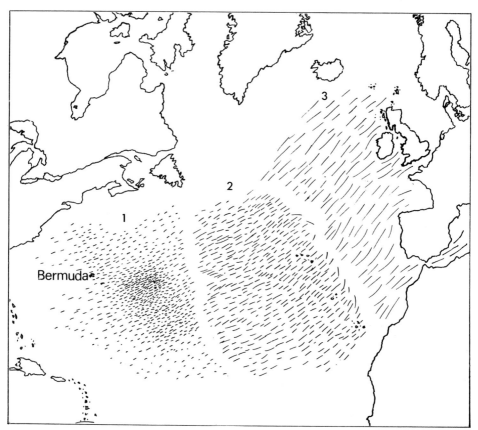

36 **Drift of eels.** The migration of the adults has not been observed but is inferred, as the opposite to the drift of larvae. The latter, the drift of the so-called 'leptocephali', was traced particularly by Johannes Schmidt. Schmidt found the smallest larvae south-east of Bermuda; as he proceeded northwest in the 'North Atlantic Drift', larvae became fewer and larger. Schmidt divided the data into 3 geographical groups. In the Bermuda to Sargasso region any 1000 larvae included one only longer than 20 millimetres; in mid-Atlantic 582; and in the European approaches *all* of them were of this big size. I take the data from Bertin (1956).

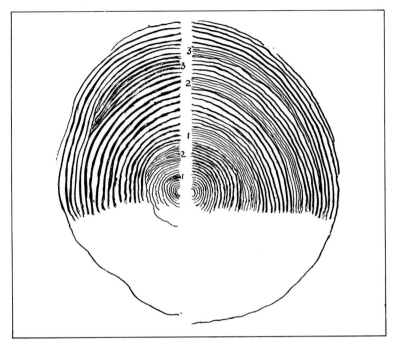

37 Migration patterns recorded on scales. Meek (1916) published this diagram drawn from photographs of, respectively, Atlantic salmon on the left, Red Salmon, a Pacific species, on the right. Both were in their fourth year. The diagram of the Atlantic salmon shows the small growth of two years in fresh water followed by rapid growth in the sea. The red salmon of the Pacific shows regular sea growth following descent, evidently in the first year.

Schmidt's data were presented in the form of contours, which have been reproduced in several books, for example in J. R. Norman's *History of Fishes* (1931). I give here Bertin's presentation. The inference from the chart (Fig. 36) is that the larvae were travelling on the North Atlantic Drift, which would take about two and a half years to bring them to the coasts of Europe.

We turn now to a more prestigious fish, the salmon. We could be watching a stream no more than a dozen feet wide and about two feet deep. Something is happening: from time to time the water is lashed and churned and there is the swift show of a dark curved body as the salmon pairs work the gravel into troughs, sow it with eggs and milt and cover them with mounds of gravel called redds. There the eggs may be held fast in the gravel in spite of the current, and will stay damp even if the stream dries to a trickle. And if the water remains deep, the eggs will still be well aerated, being at a place where the current keeps the gravel clean. Alexander

Meek in his book *Migrations of Fish* (1916) so described the spawning, which he often went to watch in the upper reaches of the River Tyne.

There are scores of salmon on the spawning ground, and if by chance you obtained a scale, perhaps one loosened in the turmoil, and put it into an old envelope and examined it later with a lens, you would most likely see a record of the life of the great fish up to that time, imprinted upon the covering of the scale itself.

The scales drawn here (Fig. 37) show, untypically, only one winter on the sea after the winters of life in the natal stream. In the stream the fish was called a 'parr', was brownish in colour and in appearance generally something like a brook trout. Most of the salmon on the spawning bed have had two or more winters in the sea, which they first reached in the 'smolt' stage, that is, they were larger than most of the parr, but especially more silvery. The scores of salmon that have returned to breed are the survivors of thousands of smolts which dropped down the river into the tidal water, and most of which were never seen again. It used to be thought the salmon's sea life consisted of hanging about in coastal waters near their home river, to which they usually return, but in recent decades some fish tagged in the sea off Greenland and others in the Baltic Sea have been retaken in rivers of New Brunswick or of Norway, far away from the place of tagging. Henry Williamson, in his animal story book *Salar the Salmon*, made the hero range the North Atlantic, and such an active and strong and urge-ridden animal as a salmon seems not likely to spend its sea life in hanging about coasts, with lumpsuckers and gobies. The work of an author of fiction has proved more nearly right than the unwarranted assumption of less imaginative students.

There has been much controversy about this sea life of the salmon and few people are willing to subscribe to any standard account of how the salmon behaves in the sea and particularly of how it finds its own river again as most salmon seem to. I have had the advantage of discussing the subject with my colleague Dr. Dorothy Witcomb and have also been in correspondence with my old friend Professor A. G. Huntsman of Toronto. I find it possible to set down a rather simple account of what may usually happen in the sea life of the salmon; but it has some gaps and uncertainties, on which Huntsman's papers are worth study.

It is usually in their second year that the smolts descend to the sea, and it is fundamental that the vast majority of them are, as already mentioned, never seen again. As they grow, some are caught in the fisheries off Greenland and in the Baltic Sea, but there is no reason to assume that they all go to those places or go anywhere else in particular. In all productive sea areas, there are millions of young and small fish that we rarely see, for salmon to feed on. For a few of them, sexual maturity comes after only a single year of sea life, when they are called grilse. With the ripening of gonads comes an urge for fresh water life, for the ascent against the current, 'as men have an urge to climb mountains', as Huntsman puts it. We have noted that the grilse run is of less importance than the run of salmon a year later, which is often repeated in succeeding years.

I do not doubt that ascending salmon taste their way home, that when encountering the flavour of their native river they turn into it, as animals generally tend to home to familiar environments. Turbulence, Huntsman writes, excites them so that they leap. He reports their jumping always against tide at the tidal rapids, the 'reversing falls' in the river at St. John, New Brunswick. Thus, reaction to current overcame direction of migration and Huntsman sees this as a key observation to the whole movement. Huntsman latterly seems to have gone further than ever. In a letter to me dated 25 April 1971, he writes: 'I here declare that salmon "migration" is a superstition, "migration" is an active not a passive verb, and salmon are carried even in shallow estuaries. There is an *aller et retour* current system'.

In the eel, the establishment of knowledge of the migration was from plankton studies; in the salmon, knowledge has been greatly helped by the readings of scales; but to some extent, as in birds, direct evidence in fishes has been obtained by attaching numbered 'tags'. Information on migration in fish species comes directly from fish marked by scientists and returned by the captors to the office indicated on the mark or tag. Some faults in the data occur as a consequence of this procedure. As with all other data, bias is much worse than bad sampling and movements shown by the return of marked fish are biassed, in that they must depend on the activity of fishermen. Scientists rely on fishermen to return the marked fish, consequently, migration to rocky or stoney grounds, or to the relatively unfishable middle waters, may not be properly recorded. Another hazard arises from a fact one soon discovers on plotting the data, that if one marks or tags fish only at Point A and several are returned from Point B, a chart of these results will suggest migration towards Point B. It is not generally recognized that in seas and great lakes where fish cannot be seen, as well as marking to test one hypothetical movement, one should also mark to test its opposite, as a control experiment in the classical scientific manner. In fact, alas, I have not heard of this precaution being taken deliberately.

Few marking experiments have given such reliable results as those made in the southern North Sea on plaice, by Garstang and an older generation of naturalists. I have been concerned in the

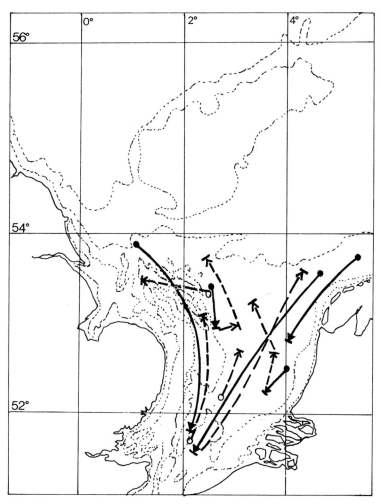

38 Migration of plaice to the Southern Bight of the North Sea. Borley's analysis of Garstang's data shows movements of plaice to the south in winter and to the north in summer (broken lines). Since all the recaptures were made by fishermen, Borley had to allow for the possibility that he might be merely recording movements of fishermen, but there is corroborative evidence in the great winter spawning in the extreme Southern Bight, mentioned in Chapter 11 below. After Graham (1956).

study of the results and the findings are available in a book that I have mentioned before called *Sea Fisheries*, usually indexed in libraries under my name as editor. I show two simplified diagrams here (Figs. 38 and 39). Mature fish were returned from two areas in particular. These areas were already known to be spawning

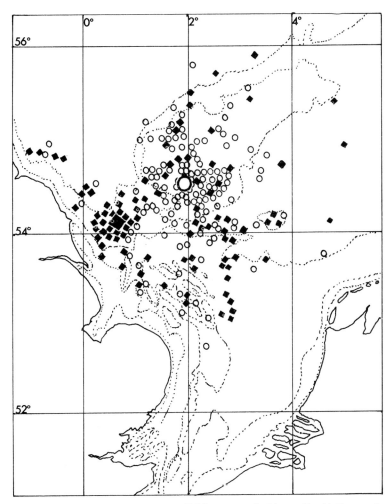

39 Plaice migration from Dogger. All fish were marked (tagged) with Petersen's buttons and liberated by Garstang and his colleagues near the circle shown on the south west patch of the Dogger Bank. The immature fish (circles) show only random movement, but the mature ones (squares) show aggregation on the Flamborough spawning ground (the large circle and square show where there were too many to plot). Each section thus serves as control for the other; neither can be due merely to movements of fishing. After Graham (1956).

grounds from the 'clouds' of eggs floating in the water over them. Immature fish were not returned from these areas and so provided the control for the experiment. Thus, to summarize the results shown in the diagram, the immature merely dispersed, while the mature were retaken in the two spawning grounds.

Similar experiments were carried out both in the summer and winter and showed that the downstream movement is the summer one, after spawning, and the upstream one is a winter movement. The fit to the seasons adds to the authenticity of the original conclusions.

Clearly there has to be concentration for mating, and then a movement of dispersal because the fish are hungry; after feeding little, if at all, during spawning.

The migration of North Sea herring is entirely composed of a concentrating and dispersing alternation. What was once thought to be a movement around the British Isles is really a wave of spawning concentrations, activating different stocks as it goes round.

The pattern of movement of the cod of the Icelandic and Norwegian grounds is up current to spawn, followed by dispersal down current, but the cod of the North Sea differs. There are four main swirls in the great square bay known as the North Sea, and there is a cod spawning ground in each; thus the fry are kept moving around until they are large enough to make bottom somewhere on one of the banks of fringing shelters, of which there are many.

Meek's law

Cod, plaice and many other fish lay pelagic eggs, that is, eggs which float freely (*pelagos* = the sea). Those which lay demersal eggs, that is eggs that are on the sea bed, usually have pelagic larvae which feed on the multifarious animals and plants of which plankton is composed. In the helpless stages of their life, that is as eggs or fry, fish are thus at the mercy of the movement of the water. This demands counteraction by adults, described by Alexander Meek's law: he pointed out that most fish when going to spawn are 'contranatant', that is, they swim against the nett current, i.e. excluding tide. After spawning there is 'denatant' movement, meaning that they drop 'downstream' to feed.

In the rough and in general, Meek's law is a valid description. Thus the spawning migration of the eel in the sea is against the North Atlantic drift, on which the larvae will return. The salmon run is upstream to spawn and, as we have seen, the plaice goes up to the entrance of the North Sea where the prevalent drift carries the fry back to the shallow nursery grounds of the continental grounds where they grow to maturity. Cod of Iceland and Norway provide further examples. Now, fish often maintain their position over the bed of a stream by swimming against the current. If, then, we go on to interpret Meek's contranatant movement as being merely another form of this counter to the current, we shall often be in error: as is fairly obvious if we consider both salmon and eel together. Meek's Law is an effect, not a process. Furthermore, current in the sea usually means not a continuous water-flow but the nett effect of tidal ebb and flow. Thirdly, unless the fish can see, or otherwise sense, the bed of the sea, its orientation in respect of the movement of the water would probably have to be by other external frames of reference: it might be the sun or stars. Nevertheless, we noted Huntsman's observation at the reversing falls, which showed salmon orientated against the actual water movement. So, Meek's Law can sometimes apply directly, but usually is best taken as expressing results, rather than actions.

In all those examples and many more, the determinant in fish migration can be distinguished as movement of water. But in swirls Meek's Law is unnecessary and the factor appears to be the necessity of concentration for mating.

Niche-filling
Whereas Meek's Law may be applicable where there is a current, something more seems to be required as a general principle. It used to be said that 'Nature abhors a vacuum'. Neal instanced this in a wood and it is a universal law in ecology; it has been recorded that even the snow-slopes and the hot sulphurous springs have their characteristic biota. It seems that all niches must be filled, and this is often done by migration. Swift and shallow upper reaches of rivers provide aerated and secure beds for eggs of the salmon, with nursery reaches conveniently not far downstream: so those beds must be used even if, as for the Pacific salmon, the spawners must travel a couple of thousand miles to find them, and die after the event. The Antarctic convergence breeds rich populations of plankton, suitable for terns, so terns must come even from the utmost part of the earth to the Arctic. Gravel beds in shallow seas are good for hatching attached fish-eggs, but as they are completely exposed to predators, they are suitable only for very numerous populations, namely the herring. The underside of leaves in Thurlbear Wood are crowded with aphids. Blue tits eat those up in the canopy, but winter limits the population of blue

tits as does the number of nesting holes, so they can not be so numerous as to deal with leaves lower down in the flush of the aphis season. Three species of leaf warblers each hunting a different level, weak and frail birds as they are, yet journey to that wood from Africa, and so distribute themselves that no suitable English wood is without them. Thus in feeding migrations, the determinant seems to be seasonal variation in the food species; but other migrations seem often to have no better determinant than the 'insurgence' as it has been called, of animate nature, which seems determined that nothing shall be wasted.

Observant fish and birds
Reading of all the migration results, the question keeps obtruding, of how the fishes and birds find their way. As we have seen, it has become known that both groups can be guided by the sun, making the appropriate allowance for the time of day. This seems incredible to the human mind, but allowance must be made for the exactness and minuteness of observation of which fishes and birds are capable. Roy Harden Jones has told me how—annoyingly to the experimenter—both fish and birds can be guided by minute scratches or pinholes on the walls of their container, whether it be an aviary or an aquarium, marks that we can hardly see when they are pointed out to us. I cannot help thinking that even the stars in the firmament are, as it were, of the same order of magnitude as the little marks in question and that a fish or bird off its course would be uncomfortable once it had acquired, as a permanency and as a normal feature of its environment, a pattern which we hardly notice as we go about our business; although we can use it if we have not too many distracting stimuli reaching our brains.

There is still a problem about the imprinting of these patterns as part of the normal make-up of the bird or fish, as a part of the environment in which it is comfortable. That problem of how such imprinting could take place, either during the life of the bird or fish or in the history of its evolution, lies, I think, outside the scope of lectures on ecology, but the student should consider adapatation as expounded in Hardy's book *The Living Stream*. He will also consider possibilities in electrical sensitivity and Selous' extra-sensory perception.

England and Africa
To return to the ecological side, I would stress again that when we treat eco-systems as if they were self-contained—when in fact they do contain migratory members as in Thurlbear Wood—we

should always remember that migration may upset our analysis. For example, the economy in some small area of South Africa could be, is, and must be affected—by anything that happens to insectivorous birds in Thurlbear Wood: to perhaps a minute extent, but perhaps a greater degree. This is a consideration that is often forgotten, and in practice we almost must begin by ignoring it. In the ordinary way populations have settled down to an average proportion of imigrants. It is usually possible to proceed by considering the local populations as self-contained, and to rely on the fact that there will always be a large number of, for example, warblers, to come and eat the aphids. It is well, however, to remember that we should be in Very Queer Street if they failed.

Migration was the first general question thrown up in recounting Neal's studies of woodland ecology; a second is the relations of predators and prey, which deserves a chapter to itself.

9   *Predators and prey*

The balance of nature
The synecology of the previous chapters has often shown a senior predator, living at the top of the food chain or near it.

Such a senior was the cod taking the herring, as mentioned in Chapter 1; in Chapter 3 the starlings had the wireworms, in Chapter 5 we saw various routes in freshwater leading to pike; in Chapter 7 there were a few senior or 'top' woodland animals: sparrowhawk, owl, badger, fox.

In the present chapter we look for any relationships of predator and prey that can be classed as general, and may be traceable through some of the numerous examples.

In the woodland, aphids support hoverfly and ladybird larvae, which support the small insectivorous birds, which also eat the aphids directly, which helped in Neal's example to support the sparrowhawks. For better understanding, let us speculate on what might happen, using the situation as Neal described it, not necessarily as it has become since sparrowhawks became rare.

Let us suppose that a gamekeeper decides to shoot the sparrow hawks, at a season when they will not be replaced, for example when the sparrowhawk population has settled down into its breeding areas. He probably hopes thereby to rear more pheasants in the wood, and may very well do so. Without sparrowhawks, we could reasonably expect the insectivorous birds also to rear their fledglings more successfully, and therefore the number of birds would be greater and the number of aphids surviving the whole season would consequently be less. This is an effect that the gamekeeper would not necessarily foresee. Fewer insects might well mean that the brambles would flourish more than usual so that when the gamekeeper came to mow brambles he would have more work, and that through shooting the sparrowhawks. We can be sure that the sequence just given would not happen every time, if at all; otherwise, by now, keepers would have learned not to shoot sparrowhawks.

In general, eco-systems are highly flexible. In 1947, there was a very hard winter in the Forest of Dean and fewer blue tits survived than usual. As a result, the flycatchers were able to occupy more nesting boxes than they had in the previous year. From synecology, we can not say in advance what effect removing one or another of the members of the web of life would have. In general, we should expect the situation to be restored. If there were fewer aphids wintering over, then there would not be so many for the insectivorous birds in the following year and consequently they would not have such a successful rearing season, and that would allow more of the early stages of the aphids to pass on to the next stages and take more sap from bramble shoots, thus tending to restore the situation. After I have been discussing these things with students, they have had good marks when they have explained the steps by which they thought that the balance would be restored, after a disturbance of the eco-system. The conception reached here is of a web of life suitable for the conditions, and therefore to a certain extent foreseeable from experience of a similar conditions. I think that experienced field naturalists would generally be able to say what species of birds or of other animals they would expect to find in particular localities, and they would often be right; but experienced field naturalists also know that sometimes they would not be right, owing to conditions overlooked or to pure sampling error.

Number control
In general, the idea of a balance of nature, with the predator having its part to play, is fairly widespread now, but this has by no means always been so.

Almost universally, in most countries throughout history, there have been bounties for killing wolves or other predators. It seems only common sense that if one wished to use the prey animal, it would be necessary to keep down animals that appeared to interfere with it. I do not doubt that in many circumstances, as when one wants to eat the sheep oneself, or as when Lapps follow and look after herds of reindeer and live upon them, it pays men to shoot wolves; that is, for all the pastoral ways of living. When wild or semi-wild conditions are aimed at, payment of bounties on wolves and mountain lions in the new countries of the Americas has been somewhat discredited.

I was very impressed to hear about this in 1951 when I had the good fortune to hear a lecture by A. S. Leopold, later a well-known Professor, speaking with authority and without doubt, 'from the spruce forests of Maine to the chaparral of California's coasts' (see Fig. 40).

When forest is felled for lumber there is a vigorous development of new brushwood. At that stage, deer populations build up rapidly. The steepest of the three curves in Leopold's diagram shows the subsequent course of events if this outburst of deer population is not checked. The mechanism seems to be that the brushwood is so severely browsed that only the swiftest growing species of trees get away towards making a canopy. All competition against those fast growers is eliminated by the deer, and so the forest goes quickly to the stage of having a canopy. When that happens the shrub layer is practically abolished and the field layer has little strength either, using the terms of Chapter 7. There is therefore little deer food. The absence of a check on the prey population, that is the deer, has in a short time proved disastrous for that very population. We are as it were illuminating the principle that may be credited to Malthus. We can agree that in order to recolonize after any disaster, and in order that there shall be no shortage of new individuals to fill any gap in the habitat or even a suitable vacant niche, the reproduction rate of nearly all animals and plants is grossly in excess of what is needed to keep a stable population. It may well be a general principle that normally the reproduction rate is countered by predators and that, if it is not so compensated, worse things befall. We have just cited Leopold's example of deer population, if unchecked after clearing forest, murdering the brushwood that should give it nourishment.

That, however, is only part of the picture. Let us suppose, as he supposes, that the deer population is not allowed to get out of control, Leopold then finds that actually a certain amount of browsing by deer of the brushwood, or as we have called it, the shrub layer, does increase the amount of deer forage within their reach. It causes the shrubs to branch, as when we clip a shrub in the garden, and, provided that these branches survive for a certain length of time, the more browsing there is at this comparatively low level, the more the food on which this comparatively low number of deer can browse. Consequently, he draws a second curve (Fig. 40) showing how the deer and their food can increase moderate dimensions, and with the canopy forming rather slowly there may be a sustained period of flourishing shrub layer. The formation of conifer canopy would mark the end of the wood as an ideal range for deer.

However, there is a third possibility: if the predators or the shooting are altogether too much, then there will never be very many deer, as in the state when the forest canopy takes over. It is evident that the authorities who paid bounties on wolves and other

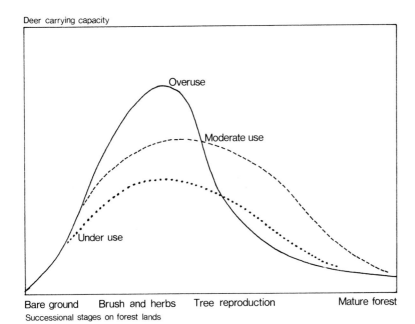

40 Leopold's concept. The three graphs represent deer carrying capacity. Leopold summarized his work leading to the conception of optimum production both of deer and timber by moderate use of both. The height of graphs represents the head of deer and the lower scale shows the corresponding stages of forest development. After Leopold (1951).

predators assumed that the deer population would always be in the state represented by the bottom curve, whereas experience has shown that the more flourishing conditions shown by the two upper curves are the ones most likely to occur, and that the effect of the predator is to make possible the condition of the intermediate curve.

The North American experience summarized by Leopold must be classed as well tested. More flourishing populations of prey result from preventing short-lived population explosions. Evidence from artificial conditions where there are no predators, supports this generalization.

A clear and interesting case is that of Crowcroft and Rowe's experiments with mice, which have been reported in *Mice All Over Us* by Peter Crowcroft. Here the conditions were artificial; there was unlimited food; there were no predators and the result was miserable for the population of mice. With no shortage of food, the ill-effect was merely due to, it seems, too little space. We are reminded of Hediger's conception, gained from various experiences including his own in the zoo at Zurich, which he directs. He has stressed what he calls the social distance: many animals, in most circumstances, do not like others to be too close to them. However, without explanations, let us look at the facts as reported by Crowcroft.

Crowcroft allowed one colony to grow from an original male and two females, and in forty-eight weeks there were a hundred and forty mice, of which sixty-seven were females. All females except one of the original ones were non-breeding. This can be detected in mice and other small animals because the female orifice closes right up and no penetration is possible. Among the hundred and forty mice, a census found no infants and no growing animals. In the same evening, when the census was finished, a door was opened allowing the mice to go into an additional pen.

Eight days later, when the population was examined, nineteen of the twenty-nine females that had gone into the new pen were breeding, and fifteen out of the thirty-eight in the old pen. The population explosion had taken place in order to colonize the new area. Until that happened, the mouse population, in the absence of predation, was prevented from breeding altogether.

The point of including Crowcroft's example in this study is to show that a world without predators may be no good world for the prey. The predator allows personal fulfilment.

It should be mentioned that in wild species of rodents, there is an adjusting mechanism, which has not been shown as connected with predators directly, but may be so connected through food supply. The mechanism is an adjustment of the litter size, that is of mortality within the womb, and of mortality in the litter thereafter, perhaps effected by the milk supply and so by the abundance of food. In this mechanism predation, if acting, would play its part indirectly, in keeping population size from outrunning food supply.

Sometimes a predator will keep a population virtually invisible, as a good cat does on my farmlet, yet catches enough mice to make meat feeding unnecessary. A conspicuous example is that of the limpets, as demonstrated at Port St. Mary, noted in Chapter 5. Huntsman in 1941 published in the *Journal of the Fishery Research Board* in Canada, positive effects on salmon populations of controlling merganzers and kingfishers of the Margaree river, but only for one year's observations.

Therapy by selective predation: territory

Although there can be little doubt that predators help the prey by control of sheer numbers, there is also a more intense effect in

Predators and prey 129

which the habits of the prey direct the action of the predators towards the less healthy members of their company.

In 1928, in Africa, I went shooting, which I do so badly as usually to cause my companions some amusement. I did amuse them also on this occasion, although the shooting itself was above my usual average. After a weary day's walk across the plains, I saw afar off a herd of Thompson's gazelles, which are pretty gazelles

41 *Morrow's unicorns.* Unicorns will serve to illustrate the great number of herd animal species in which a dominant member has his choice of the best grazing, with the rest of the herd members in ranking order grazing warily around. By this arrangement, the weaklings are at the more dangerous outer ring, which illustrates the therapeutic value of even temporary territorial behaviour—which is how the grazing order may be described. From Graham (1956).

with a black stripe on the flank against a background of white. I did not get very near to the herd, but there was a lone gazelle that stood between me and the main body, and I was able to get close enough to take a long shot, which I am glad to say killed the gazelle. With field glasses, I had inspected the animal in profile and seen a big horn and I thought I would have a nice trophy. It turned out that the horn I had seen was the ram gazelle's only one, which was doubtless why it was distant from the herd, having been driven out, and so, as a solitary had become an easier animal to stalk: whereas in herds there is usually one or more with its head up, and so watching. In George Morrow's picture (Fig. 41) the leading unicorn stallion, strong and healthy, is central and on the best grazing, whereas on the outskirts are less healthy outcasts. I chose unicorns rather than any living animal, because of uncertainty between matriarchy and patriarchy. Frazer Darling in *A Herd of Red Deer*, found matriarchy, making 'the monarch of

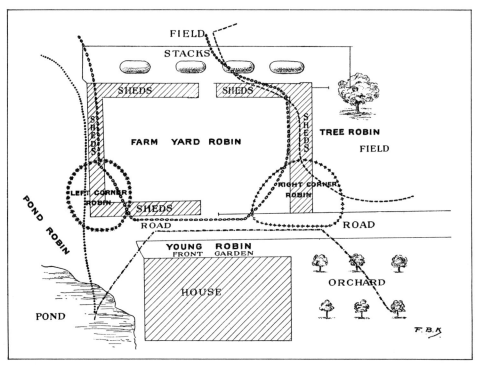

42 **Kirkman's robins.** In winter, Redbreasts are solitary, as has been known from earliest times. F. P. Kirkman mapped their 'estates', now called 'territories' and wrote about them. Reproduced from *The British Bird Book* (1911) which he edited, by kind permission of Messrs. Thos. Nelson and Sons.

the glen' a mere hanger-on. To return to the picture, the centrifugal arrangement, whether of males or females, provided therapy by predators. Thus there was therapy, by the herd placing the one-horned animal in the way of me as predator.

I have argued in my paper in *Human Relations* that this hierarchical, or pecking order, division of grazing, is at one with territorial behaviour, which has been known to naturalists for much longer. Kirkman describes the territories of robins that he kept under observation at a farm (Fig. 42). The robins' division is hierarchical, the order established by song and by squabbling, and results in mating and breeding areas. But it also has to do with feeding, as in the grazing unicorns, because the robin parents need an abundance of insect food near their nestlings, and the competition for best sites is also for best hunting ground. It is further highly probable that the lower robins in the pecking order will have to put up with less cover in the territory, and therefore be more vulnerable to predators.

More recently, Wynne Edwards has written a fine book on territorial behaviour and related problems, *Animal Dispersion in Relation to Social Behaviour,* which tells much more than can go in this chapter.

However, I think that almost the clearest case of territorial behaviour is in fish. It was worked out by G. P. Baerends and his wife, Baerends van Roon. During the occupation of Holland by the Germans, when field observations were hampered, these young naturalists made good use of their restricted area of activity by studying the behaviour of tropical fish in aquaria and thereby advanced knowledge of the whole subject of territorial behaviour more than they might have done in many months of safari after larger animals. It is important to notice what happened (Figs. 43–6). As the little male cichlids became highly coloured and sexually mature, they tended to establish territories on the bottom of the aquarium tank. The first one to colour up claimed the whole tank floor and chased any other away from it, but as other males came into breeding condition, he had to make room. By continual squabbling and display on the border, they made their territories. When a male is fairly on his territory he always succeeds in winning, but there are continual border disputes, as each one tries to enlarge his territory a little. Which one wins depends on how near they are to the centre of the territory. The invader must not go far in, or he will take a beating of no mean dimensions. The birds and the field mice do the same, and I think I am right in saying that almost every vertebrate animal does so. Certainly it has been observed in a good many. Men and domestic dogs are worth noticing in this respect.

Among the Baerends' fish, there were some who never achieved territories. They lived, chivvied about, in the upper part of the tank, and never succeeded in breeding.

When near the surface, fish are visible to predators from above and below. It is, I think, fairly clear that the same happens in many wild animals, where naturally the phenomenon is not so obviously severe because they are not in a confined space. I have always been

---

43–46 Territorial behaviour in animals. Some examples of the observations of G. P. Baerends and Baerends van Roon on *Hemichromis bimaculatus.*   *43   Social contacts. Above:* A trespasser and an occupant go circling in lateral display. Each makes himself look as large as possible and is flushed with red, and carries metallic blue patches, shown here by white blotches. He lashes his tail to direct water at the other fish. *Below:* the trespasser has been out-faced, and withdraws, seeming now as small as possible and with protective coloration resumed.   *44   Interloper.* The Barendeses wrote:

A tank of 60×30×30 cm contains 10 individuals of this species: as later appeared to us, 1 male and 9 females. They can be distinguished individually by different coloured threads attached to dorsal fins. To be able to indicate the position of the fishes the area is divided in 2 layers of imaginary compartments, numbered 1–8 in the lower and 1a–8a in the upper one. . . . Plants are equally distributed throughout the aquarium and in 3 lies a flower pot on its side.

Now we come to Dr. Stanley Frost's arrangement (*above*) in order to draw the whole aquarium and contents on one plan. He imagined the upper 'floor' to be opened and folded flat. He then represented the fish as circular, so that their vigour and strength of colour could be shown by relative size. The events of October 20th can be shown (*below*).

October 20th: Hemichromis. A possesses 1, 2, 1a and 2a and frequently visits 5, 6, 5a and 6a. B possesses 4, 3, 4a and 3a and frequently visits 7, 8, 7a and 8a. The rest of the troop is usually found in 6, 7, 6a or 7a. Only C is often seen in 3a where it tries to drive a wedge downwards. It is chased by B especially, in whose area it chiefly operates, but after each defeat C returns. 14.30: C now manages to hide itself in the flower pot. 15.00: C leaves the flower pot and swims to and fro for some time in the middle of the tank near the bottom. When it is attacked by B it does not retreat, and B goes back again. 15.15: C is in the pot again, B appears and tries to get in, C displays frontally, then B turns pale and disappears.

45   New claim succeeds

October 22nd [above]: A and B still have the same territories and C now steadily maintains its place just outside the flower pot where it has boundary fights with A and B. When attacked, C often makes the typical maintaining attitude with the tail pressed on the bottom. Whereas on the 20th of October C always fled into the pot when a neighbour approached, it now behaves aggressively, in particular at about the line that separates 2 from 3, which is the boundary between the territories of A and B, and therefore where resistance is less than at other places. C also chases territoryless fishes. Its red colour has become brighter and is already nearly equal to that of A and B. In 6 and 7 another fish, D, somewhat brighter-coloured than the other fishes of the school, tries to become dominant. A similar fish E has established itself in 6a, and sometimes fights frontally with D. A third fish F is often seen in 7a: it never puts up a fight but regularly returns to this area after having been chased. A and B often swim in the territories of D and E without being attacked by the owners. When B goes to 7 or 8, C often takes this opportunity to enter 3 or 4.

October 23rd [below]: A, B and C still have their territories, and are equally bright-coloured now. The territory of C covers a part of the former territories of A and B. D and E are not so brightly coloured, although the markings of E as well as those of F are brighter red now than yesterday. Today F fights other fishes. D, E and F fight each other frequently, but only defend their territories weakly against A and B. The other fishes (H, J, K) mostly stay in the neighbourhood of D's territory. They now and then chase each other.

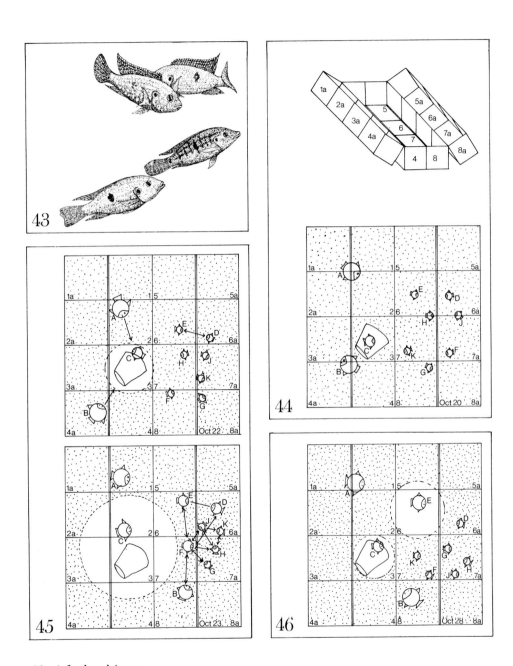

46 A further claimant

October 28th. A and B maintain their territories, and C weakly tries to recover the lost area. E is fiercely defending its territory, but the other fishes are less active.

Reproduced from Baerends and Baerends van Roon, *Ethology of Cichlid Fishes* (1950), by kind permission of the authors and E. J. Brill.

astonished to notice that when a bird loses a mate, it is not very long before another mate turns up and I can only believe that it comes from this pool of unsuccessful members. During the breeding season we will still find moorland birds on the sea shore. They, I assume, also correspond to Baerends' homeless fish. I thought that it was a legitimate inference, when I proposed the unicorns, that thoroughly low members of the pecking order, for that is what they are, are forced into more vulnerable positions and fall more easily as prey to predators. Morrow drew them as 'screws' (Fig. 41). Lately, however, we have definite proof from Peter Crowcroft's mice. The lowest members of this community were forced by the competition of the stronger or more vigorous ones to feed during the daytime, whereas it is obviously more natural and therefore we can assume safer, for a mouse to feed at night.

Emigration
On the subject of emigration and lemmings, etc. I used A. S. Heape's book *Migration, Emigration and Nomadism*, published in 1931. There have been some papers since, but to my mind they have not added very much to the integration that I gave in 1964 in my paper in *Human Relations*, which was based on Heape's, with the addition of something on crowd psychology, including territorial behaviour. I will summarize it briefly here.

The lemming is a small rodent, something like a hamster, that lives in Norway. The animals often achieve a greater population than the land can support and thousands of lemmings start walking towards the westward because all the best land, which is towards the eastwards, has been taken up. The phenomenon starts with one or more years of very successful breeding and it seems to me that these who start walking correspond to Baerends' fish at the top of the tank, constantly driven to and fro. Having started walking, I have little doubt that crowd psychology takes over and that they all feel a compulsive movement to walk to the westward. They are terribly preyed upon as they go, and finally those that survive the journey perish by plunging into the sea. It is not a pleasant phenomenon at all. Heape thought that lemmings had to go because of stain on the grazing, which we mentioned in Chapter 3, as might well be. Of course, my integration has no certainty, but for the time being it seems sufficient to regard emigration as a piece of territorial behaviour, with the banishment of those who are unsuccessful in the struggle for territory.

Mr. Zweers of Amsterdam has told me of great tits squeezed

out of a detached Frisian island periodically having an emigration, but in a year of emigration too many leave and the population has to build up again, this seems to be crowd psychology (Le Bon). These are H. N. Kluvers' research results.

Parasitism

Both Professor Kershaw and another colleague, Dr. Boris Fistein, have mentioned to me that the parasite of sleeping sickness, the so-called trypanosome, probably keeps the level of animals down to a reasonable number for the ground to hold. This, I learn from Professor Kershaw, was an insight of the late Patrick Buxton. In Africa trypanosome infects cattle and antelopes, and in South Africa, small rodents. The parasites are reduced to quite small numbers in the animals that survive an initial attack, and it is thus suggested that the parasites serve the prey population in the way that we have seen the predators doing, in restricting numbers. Professor Kershaw and colleagues at Liverpool conducted experiments showing that rats carrying a small number of parasites, which is counted in practical life as immunity, are not in tests less intelligent, nor do they seem to suffer very much in any way, compared with uninfected ones. The results are reported in the *Annals of Tropical Medicine* from 1959 to 1962. It would therefore seem that it is possible for a parasite to benefit the prey animal if there is no predator to do so. However, this question is contentious and there are many different circumstances. I have lately come across one local case in which the infected prey animal dies and the predator is normally unharmed by the parasite, which is a tapeworm. Dogs and cats do seem to be able to live on terms with tapeworms. I live near a public wild park and my neighbour brought me the cyst that he had taken from the brain of a sheep which was running in circles. When I brought it into college, Dr. Phyllis Wells showed me that it is *Taenia multiceps,* one in which some hundreds of small tapeworm larvae stay in the brain, folded in on themselves so that there is no structure to be seen until you put them into, for example, saline solution, when they evert, that is turn themselves right way out and reveal the head with its hooks and three segments of the tapeworm. If a dog or wolf had eaten the brain of that sheep it would have acquired a tapeworm. The segments of that tapeworm would eventually fall on the grazing and the eggs and larvae would re-infect another sheep. But let us note that the cyst, by occurring in only one of the lateral ventricles of the brain, caused the sheep to run in circles, so that there would be no difficulty for the predator in running it down.

That is an example of the intervention of a third party, the parasite, in the predator-prey relationship. In a sense the parasite is predatory both on the wolf-predator and the sheep-prey. It is one of the facts of parasitic life that the numbers of its prey are very rarely much affected for example, in this case my neighbour has suffered to date only four cases in four years, out of a standing head of adult sheep of two hundred and fifty to three hundred. However, as is well-known, that is not an invariable rule and epidemics do occur, leaving perhaps relatively few survivors. This contentious and variable subject is included in this chapter in order to point out that it is not necessary to believe that a parasite is always harmful to the stock on which it preys. It may help as much as a predator does.

Human reactions

Human nature is essentially pitiful. Predators arouse a kind of hate in us; we can hardly bear to see the lioness knocking down the zebra, the cat catching a mouse or a magpie persecuting and killing young birds or worrying the older ones by stealing eggs. In all these cases we are tempted to interfere, but the evidence seems to suggest that we should be quite wrong to do so, even from the point of kindness to the prey animals. In the matter of paying bounties the question seriously arises whether it could have been kinder to the prey as well as more rational, to leave the wolves alone. I would cite Sally Carrighar on that, especially one observation described in her book *Wild Heritage*. She watched a small pack of wolves, half a dozen or so, of which one large grey wolf seemed undoubtedly to be the leader. They were following a pack of caribou, that is wild reindeer, which could perfectly well see the wolves following them but took no notice of them at all, nor did the wolves appear to take any notice of the caribou, merely following them. But, as they travelled, it happened that a calf fell behind the rest of the herd. Seeing that, the big wolf leapt into action, galloped after the calf, knocked it down and killed it and the other wolves came up. It was all over quickly. If we are to compare that with the distress that would have been caused by the calf following on feeling ill and weak but just able to keep up with the herd, and the trouble to the reindeer parent to jolly along a sickly calf—if indeed it would do so—we can reasonably ask which in the end is more merciful: the action of the wolves or the prolonged illness and unhappiness of the calf.

We may consider the example of the moose on the Isle Royale in the Gulf of St. Lawrence. There is a wolf-pack there which often brings a moose to bay, but if the moose is vigorous and shows fight the wolves do not attack it, fearing no doubt the punishment that they would receive from the hooves of the moose's forefeet. If, however, the moose does not show fight, then the wolves execute a little dance of triumph, and running round the moose they finally rush in and kill it. The whole arrangement brings swift death to the old or sick. There are other examples which lead to the same general conclusion. It can not be comfortable to be a fish that is not on an even keel and show the 'flash' of a silvery side as it wobbles, so immediately becoming conspicuous to a predator. This might be a gannet circling over a shoal of herring or it might be a predatory fish such as a pike looking out for a shoal of roach. In either case death will be sudden.

Even if it is agreed that predation is normally merciful and is a kind of welfare work for the prey, there is one group of actions that arouses a special human revulsion. This group includes the play of cat with mouse, which represents a practice common among carnivores, that of death deferred. This play is not entirely heedless, because the cat will carry a mouse, or even a bird, to some corner whence it cannot easily escape, but I agree with Sally Carrighar that the cat is not likely to be aware of anything but its own side of the game. How should the cat know what the mouse feels? Indeed, how can we ourselves know? There are, it is true, a few facts to suggest that the play is not so bad for the mouse as we might judge from merely watching. A classic fact is that Dr. Livingstone felt no pain when mauled by a lion. He experienced only numbness. Another fact is that I have seen my cat embrace a young rabbit, while biting and renewing its bite in the throat, presumably drinking blood. If so, keeping the prey alive and moving in the play may help to extract the blood. On the prey's side, reduction of the blood supply to the brain would not be expected to leave the animal as fully sensitive to its predicament as normally, but it is capable of simple actions. When I tried to catch a young rabbit in this state and kill it quickly in order to spare my grand-daughter's feelings, it could run and scream. It might have been more merciful on my part to have let the cat's game continue, quietly as it was proceeding. There is more in this matter than reaches the eye.

Here I would mention an incident related by our truly reliable ships' storekeeper at Lowestoft, the late Sam Strowger. He was watching a rat that was being played with by the cat, and saw it break off to have a drink of water, after which the play was resumed.

Deferred killing is not always play. Marais' baboon troop was preyed upon by a leopard, which would catch a straggler in the evening before the troop reached the safety of the night-sleeping ledge, but in the remains of daylight the leopard did not wait to kill it and carry the burden away, for in that light members of the troop would rally and hurt or even kill the leopard. But in darkness the baboons could not see to do so, and must stay on their safe ledge. The leopard met the situation by paralysing the baboon by biting it, and the distress cries of the immobilized baboon continued until the leopard returned for it in darkness.

I cannot further comment on this subject, but I trust and firmly believe that circumstances would usually make deferred killing rare compared with death that is swift and merciful.

In general, we may agree that Nature is not so cruel as she first appears and that the predator is, on the whole, a merciful agent. The lion is not the savager of the herd of zebras but is welfare officer. It seems to follow that human beings have to correct some of their common attitudes. One thing is quite clear—in return for all the advantages and pleasure we get from animals, we ought to give them a quick, early and merciful death.

Major fluctuations

To return to the question of numbers, there remains one subject which has become slightly awkward over the years. When I was a student it was thought to be clear. It was very interesting to hear that there was an oscillation, a rhythm, in the numbers of prey and predator. For example, the arctic hare, which is called the varying hare in North America, seems to have a rhythm of abundance of about 9·7 years. In the best years, the skins handed in to the Hudson's Bay posts may be ten times the number offered in a quiet year. Equally, the lemmings seem to have a rhythm of about five years in abundance and voles have a rhythm of three years. It used to be said that these were followed by rhythms of the predators: for instance that the numbers of lynx skins handed in in the Arctic also had the rhythm of 9·7 years and that this was probably the result of abundance of the hares; and also the cause of their reduction. So, there were interlocking fluctuations, to be expressed mathematically: convenient presentation of the data may be found in J. Z. Young's well-known *The Life of Vertebrates* and he also shows other great fluctuations, though not necessarily rhythmical. For example the Norwegian herring hatched in 1904 predominated in the catches off the coast of Norway until 1919, which was the first year in which any other hatching showed a large contribution to the total (Fig. 47). Similar effects are to be seen in

North Sea haddock, which Young shows. There is no periodicity claimed for those fluctuations, but it is to be noted that there are those very large fluctuations in fisheries as well as in rodents.

Returning to the rodent and the lynx, the relationship between them does not seem to be borne out by Young's diagram. According to Machulich's data, which Young uses, the peaks of lynx abundance seem to have slightly antedated the rhythm in the hares, more often than not. The question arises of the origin of these cycles and a trapper in Canada told me that it was due to the hares having a cycle of venereal disease. We had been short of meat for several days, as I had not even seen a moose let alone shot one, and I asked him to shoot one of the hares with his ·22

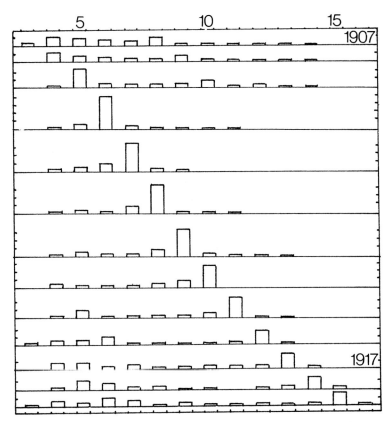

47 *Fluctuation in herring.* Numbers of many species have great annual fluctuations. This example is a classic, for which no single explanatory factor is known. It is likely to be a combination of favourable variations in the pelagic phase. After Graham (1956).

rifle, but this he refused to do on the grounds that they were liable to have venereal disease. I discounted his belief at the time, but have since found a reference to it in the scientific literature, which shows how rash it was of me to underrate him. The untutored naturalist on the spot is often a person to whom the greatest attention should be paid; probably he has not had his brain addled by too much learning, and his observations are nearly always reliable. His explanations, though, may need critical appraisal. My guide had an aversion to shooting a tree porcupine and to cooking it, refusing to join me in eating it; but he had no explanation for his unwarranted prejudice against this animal, which tasted like hare.

There seems to be no explanation for the large fluctuations in fisheries in the dominance of any particular factor, although a great many have been investigated. I think that a possible explanation is a statistical one, that in these very abundant years a great many contributory causes towards high survival all came positive. I used to say sometimes that the only reason that the North Sea does not become solid with fish is that it has not happened yet. I shall not pursue that argument, but I think that it may be that occasionally there will be a combination of factors that will open the door to greater numbers than usual. In so far as the regular rhythms are established for rodents, and I think that at least some of them are, there must be factors either in the maturity of the rodent or of some disease or parasite producing the regularity. Let us leave it at that, indeed the subject is unsatisfactory.

From this point in my course of lectures I began to draw more freely on my experience of actual ecological jobs, starting with my work on the Victoria Nyanza, in the next Chapter.

A tropical Great Lake 10

Speculative preparations
In 1927 I went to the Victoria Nyanza about one of the fisheries there. First, having no information, I made up a questionnaire asking for what I seemed to miss—the history of the fishery and native names of species and asking for a ship of a certain size and freeboard and so on. In the end everything worked, mainly because the man in Kenya who had initiated the enquiry, C. M. Dobbs, who was a Provincial Commissioner in Kisumu, took care that my questionnaire was passed around, and so by the time that I arrived there were about half a dozen very good memoranda prepared, based on local knowledge of everything that had happened concerning the fishery—especially Dobbs' own research, which was later published by the Natural History Society in Kenya. The adequate ship proved essential (Fig. 48).

History
It made all the difference when we learned that one species was paramount in the commerce. This was a fish called 'ngege' which is something like a carp in appearance but is actually of the great perch division of fish (Fig. 49). It was a cichlid, but much larger than the many species of tropical aquaria, such as the Baerends' used in their study, described in the last chapter. The ngege grew to about one and a quarter pounds. It was, we found, very good to eat, having flesh like something between haddock and turbot. This was our staple diet, because meat was scarce; this fish we ate three times a day for the seven months that I was on the lake. After the survey it was scientifically named *Tilapia esculenta*, Graham.

The trouble reported in the ngege fishery was thus: from time immemorial there had been fishing by indigenous methods such as basket traps (Fig. 50), until 1905, when a Norwegian named Aarup introduced long flax gill nets which were hung in the water (Fig. 51). In these early days the catch was about 25 fish per 100 yards of net. The fishery grew and in 1921 about 20,000 such nets had been imported from Ireland. The progress of the fishery, which was virtually in one area of about 10 miles from Kisumu can be

48 The fishing survey of Lake Victoria, 1927–8. Illustrating: a freshwater lake as big as a sea; a research ship the S. S. *Kavirondo*; a Buganda type of canoe; and reeds and papyrus—the latter prominent in the ecology of *Tilapia esculenta*, Graham.

judged from the licence fees taken at Kisumu. Between 1917 and 1921 the fishery flourished; then there was a levelling out period from 1922 to 1927 when the catches were approximately halved. In 1927 the catches near Kisumu fell to about 5 per net and quite naturally people wondered whether there should be legislation to avoid further depletion.

A detailed history of the expedition is contained in my report

49 The staple fish. This is the ngege, the aforesaid *Tilapia esculenta*, in the condition found under a modern net fishery, that is, with numbers a quarter or less of the maximum when lightly fished.

50 Local fishing. Papyrus, *Cyperus papyrus,* arranged as a maze, with non-return basket at several key positions. It is visited on a papyrus punt.

The Victoria Nyanza which was published by the Crown Agents for the Colonies in 1929.

Survey

My jobs always seem to start with a period of soaking myself in the environment for a time, with a very open mind, into which flow all kinds of impressions, including the observations of the local people. In ecology it is essential to consult the local people. 'Do you know that there are lizards near Wigan?' I did not have time to follow it up, but the children of Platt Bridge say that they are on the railway cleaners' dump on Ince Moss. Africa, however, had its own flavour, which was not lost in local differences. Even the larger fauna show a unity: elephants, giraffe, zebra, buffalo, range from the Cape to the Sahara with crocodile and hippo in the rivers and lakes. And everywhere there were kind, cheerful, docile people whose forbears had been slaves since pre-dynastic times, the prey of raiding tribes. E. B. Worthington's *Inland Waters of Africa,* gives a good account of all this and more.

We had an English captain and Chief Engineer, a Goan second engineer, an Indian mate, a Swahili bosun from the coast and a mixed crew; with firemen from Uganda who were quite expert mechanically; and Kavirondo sailors, who, as I have already indicated, were excellent fishermen; a Kikuyu and a Tanganyikan valet. The lake men are naturally canoe men, and had to sing as

51 Net fishery. This was introduced by a Norwegian named Aarup in 1905, and it provides a basis for a commercial fishery in the Kavirondo Gulf, estimated at 1250 fish per night in 1927 at a rate of about 5 per net, as well as numerous less organized fisheries dotted about the lake shore.

they rowed our ship's boat. We used to row with them for pleasure and exercise, but the difficulty was that they could not keep time unless they sang, but we could not easily row in time with them, because the rhythm was syncopated. E. B. Worthington came as my voluntary assistant, multiplying my efforts by more than two. The thirty people I have mentioned made a friendly working space for me.

Lake Victoria is right on the equator, where the sun rises at about six in the morning, passes unclouded through the zenith and sets at about six in the evening with only some twenty minutes variation throughout the year. That may sound like Heaven to us in England, but it is not so—one misses the seasons. The lake has

A tropical Great Lake

52 Local fishing station. This is a (polygamous) family fishing station, using European flax nets, drying the ngege and sending them to trade in a dhow, which visits from time to time. It is 40 miles from the railhead and ngege density was 7 per net per night.

an area of 26,000 square miles, about the size of Scotland, and is the second largest freshwater lake in the world, Lake Superior exceeding it.

Within a couple of weeks we were afloat—still letting impressions come in—on our first voyage around the lake. That voyage took six weeks. We noticed several types of scenery on the shores of the lake (Figs. 52-5) and this proved important for our mission.

53 Scenery in ngege country. Ngege are associated with a dry type of scenery, predominantly at the south and east sides of the lake.

54 Scenery where ngege were uncommon. These are banana and cotton plantations in Uganda on the northern side of the lake, where rainfall is higher.

In the Sesse Islands the forests come right down to the shore, so sheltering the tsetse flies which carry sleeping sickness from the blood of the sitatunga antelope to man. For this reason no one lived on those islands at that time. In contrast, there was Kavirondo country—dry grass and native villages (Fig. 53). This country is dotted with hamlets, each one self-contained and self-regulating.

55 Exposed coast. This is on the west side of the lake, where long-lines were used to catch predatory species of catfish and some others. This is not ngege country. The evidence of Figs. 49–55 called for an explanation. It seemed to us, and to our sailor colleagues, that the distribution of ngege might have something to do with the prevailing winds (see Fig. 56).

Where there is shelter around the shore one finds papyrus, which is the species called in the Bible 'bullrushes', in which Moses was found, and from which Egyptians made paper, whereas where the coastline is more exposed there are ambatch trees, which produce a very light wood similar to balsa wood, and hanging in the trees are the nests of weaver birds. At an extremely exposed place on the shore called Bukoba there were actually waves breaking on a rocky shore (Fig. 55) and that region is moderately wet country. Later we understood that differences in the distribution of ngege were related to the scenery.

Around the lake there were at least a dozen and perhaps more, peoples, tribes or sub-tribes who had different names and spoke different dialects. I conclude that many were probably only dialects, because the names for the fish fell into three basic groups, that is, there tended to be at the most three names for the same fish, as I found by showing specimens. The Kavirondo of Kenya were able to distinguish species that agreed to a large extent with the distinctions made by the British Museum, of fish which had been collected some twenty years before. The Jaluo section of the Kavirondo—to be precise—were able to distinguish some of the finer points of difference between species that the other tribesmen could not.

Fishing

On the first survey our task was to observe and take samples all round the lake. It was impossible to take samples of everything, but the local memoranda and the guidance I received made it clear to me that the first priority was to take samples of the ngege. However, I soon found that ngege were not everywhere, but were confined to soft grounds.

The main sampling gear consisted of nets of the commercial type, that is, 100 yards flax nets mounted in to 70 yards, corked and weighted to hang vertically in the water overnight.

Drift and wind

As well as actual fishing, we also threw out drift-bottles which contained cards with 'return the bottle' in Swahili inside them. Swahili was the slave traders' language, it extends right across Central Africa and is a mixture of Arabic and Bantu. The late Richard Dent from the Kenya Game Department came as adviser and interpreter to us, but after about a month we had learned to speak some Swahili, which was absolutely essential. One language was sufficient because in any village there could generally be found

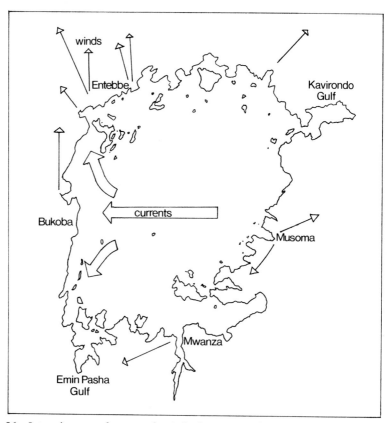

56 *Intermittent south-east trade wind.* The Kenya and Uganda Railway Marine piermasters had kindly taken for us wind direction and strength, morning, noon and night. From this data, as I had been taught at Lowestoft, I calculated vectors, taking the average for each fortnight during the survey. The second fortnight in October, which I have used here, seemed reasonably typical and I plotted the strongest of the three daily averages for each port. They do seem to show that the south-east trade wind is the most common one for the lake. We also threw out drift-bottles on a line across the water from Musoma and when their movements were processed for the corresponding fortnights, they seemed to confirm the direction of the winds. We had then several pieces of ecology of the ngege, and the question now was whether they would fit together satisfactorily.

a Swahili-speaking inhabitant.

The drift-bottles tended to drift from east to west and the men who picked them up were happy to walk perhaps fifteen miles to the District Office, carrying the bottles on their heads (see Fig. 56). In the more primitive areas such a trip was like a holiday, because anyone going into a village was welcomed and entertained as a matter of course. It was not so in the more sophisticated areas, parts of Uganda for example, which was largely a feudal, hereditary kingdom. When I was away in a far port in Tanganyika it was necessary to send a letter back to Kisumu, which was about two hundred miles away. There was no difficulty. I gave the letter to a member of the crew and off he went. There was no need for him to know the way because people would tell him, and as for lions he showed me his long sharp stick and laughingly assured me that it would be adequate protection.

To return to my primary occupation on the lake, the study of the ngege, the specimens were puzzling. The ngege could not be identified with the Museum description and I sent a specimen home. It turned out that the commercial species, which all the fuss was about, was unknown to science! It is now known as *Tilapia esculenta,* Graham, because I wrote the first taxonomic description of it in the *Annals and Magazine of Natural History,* when I came home.

Fixed station

As a result of the survey we found that the catches of ngege were localized and we were able to make a chart of them, showing positive and negative bays and channels.

After that, I did not know what to do next, as I could not see any way of solving the problem of what had happened to the fishery.

Having made no visible progress using methods that would be classed as extensive, I thought it might pay to change over to intensive study of distribution of the ngege near one spot, with all the local conditions and factors associated. Thus I would at least be building on what knowledge we had gained. We chose a place close to the mouth of the Kavirondo Gulf, in fact near Sukuri Island in the gulf where there was a good variety of conditions in one area. There was also a local fishery there, organized by a Ugandan, Onyanga, who had been adopted by the Jaluo tribe. He had three wives and, after seeing the general misery on the face of the oldest, first, wife, I have never been able to see anything good about polygamy. Wynne Edwards (1962) argues that polygamy reduces the population, which I had not thought of. However, since Onyanga had only three children by three wives, and in doing so had deprived two other men of wives, Wynne-Edwards' point is borne out. It may often be so.

Off Sukuri we anchored the ship for two months and went fishing and observing on a line of stations across a channel, with one exposed side and one sheltered. That was our intensive study. From it we were eventually able to draw up the section shown,

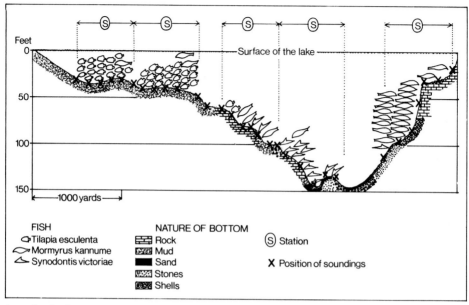

57 The fixed station. After a search lasting four weeks, a locality was found where many of the lake's conditions seemed to be represented within a manageable compass—a 'microcosm'. There the ship was anchored, and we carried out a ten-week study of *Tilapia esculenta*, the commercial species, or ngege. The species was found to feed upon what is now called periphyton, i.e. plant microscopic plankton that had settled on underwater structures or on the mud, where the water was sheltered. That is what the section showed, as can be seen in this diagram. We had thus a hypothesis to be tested in further travels around the lake.

much simplified, in Fig. 57. After ten weeks of investigation we had established that on the exposed side of the channel there were no ngege, but virtually only *Mormyrus* and *Synodontis*. Where the mud had settled there were very good catches of ngege and few *Mormyrus*. Where the currents kept the channel clean there were again *Mormyrus* and *Synodontis* but few ngege. From this we ascertained the ngege must be in sheltered water not exceeding fifty feet in depth. We were glad to have defined an habitat, but we could not see how this could help us—after all I was not supposed to be collecting elegant ecological knowledge of that kind; I was there to try to find out what to do about the fishery, and, despite all that I had learned, no answer had presented itself to that question.

Testing an hypothesis

There was then about six weeks left, and a fortnight of this was to be left for writing a report, although, as I have made plain, there was nothing of substance at that time to report on. I wished that I had been able to see and handle the actual fish of the years when the net fishery was first developed. In 1920 fishing had run at 20 per net or more. But I did not have such data; there were no sizes of fish to make even a mock mortality table, only the knowledge that the catches were now down to 5 or 7 per net, and I could not understand why. However there was still about one month of observation possible and I felt confident that we would find a solution somehow—I mustered that unreasoning optimism which I have possessed all my life. I have always guarded against being caught unprepared by good luck. In the absence of any stronger claim on our time, this precious month could be spent in establishing our discoveries so far, in going around the lake to verify or refute the hypothesis about the distribution of the ngege, which we had made from the intensive study. The hypothesis now had several strings to it. The drift bottles had all landed up on the west side, which was in keeping with the movement of the water under the south-east trade wind, which had shown up in piermaster's observations made for us. The fact that the shore on the east side was sheltered and that on the west side exposed was further evidence, as was the south-east dry scenery and the north-west wetter. It seemed from our section that the ngege were associated closely with the mud. Mud means sheltered water, and papyrus being, as it were, a floating turf, needs shelter too.

From our observations, we were sure that the distribution of the ngege was connected with that of the papyrus and mud, but how could we ascertain the total distribution of papyrus all around the lake? I was using all the time the charts made by Commander Whitehouse, which Professor Stanley Gardiner and I had pored over in the preparatory period. At this crisis in the last weeks of our survey, I suddenly realized that Whitehouse could not have surveyed properly that part of the coastline where the papyrus was thick, up to a mile often, because he could not have got his boat in to land; consequently the coastline there could only have been dotted on the chart—and so indeed I found that it was. Where the line was dotted would be, on our hypothesis, ngege areas. We could now pick out new possible ngege areas for a final voyage, and so we decided to go round these areas, lower our nets and compare the catches of ngege in those places where there was papyrus and those where there was not. Where we found ngege we also found the lung-fish *Protopterus*, which feeds on snails. This association was plausible, because the food of the ngege was periphyton,

152   A natural ecology

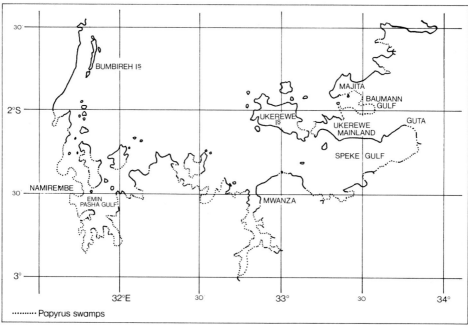

58 Victoria Nyanza, southern half: conditions. The distribution of papyrus swamps is marked on the Admiralty charts as a dotted coastline, because a surveyor's boat could not reach hard land, being prevented by papyrus mat.

the sessile diatoms in a green scum—which settled in the calm water. That was also the food of the snails and so the haunt of *Protopterus*. Whenever we took a piece of papyrus mat or sod and broke it up, we found young *Protopterus* in it, in their little chambers, and so the association of ngege and *Protopterus* in the same vicinity is reasonable enough. Comparison of Figs. 58 and 59 shows the close association of these factors. I would claim the charts as nice ecology, but this did not solve the problem.

We knew other probably relevant facts; we had been able to estimate a growth-rate from some ngege which we kept captive in a live box in the river. This was not wholly unreal because of the ngege's food being periphyton, of which plenty settled on the box. We knew the size at first maturity and the fact that ngege breeds more than once a year.

Solution
The solution thus could only come by accident or luck. When we went on in the voyage, we reached the Emin Pasha Gulf, which

is at the opposite end of the lake from the Kavirondo Gulf. I was merely testing the hypothesis about papyrus and ngege. However, there it happened: there were the large catches, of the order of 27 per net, which was the same as that reputed in the Kavirondo Gulf in 1905 at the start of the European net fisheries. This was significant because the Emin Pasha area was a tsetse fly area, and, therefore uninhabited because of the danger of sleeping sickness; the only fishing there was of the primitive kind and even so was carried out by fishermen on short visits. Immediately we noticed that the fish we caught looked different: they were large, large-headed and bony (Fig. 60). Thus, in the Emin Pasha Gulf there was a large proportion of old fish and I realized then that I had found a fish population like those of the early days elsewhere: herein lay the solution to the problem. We had here in our hands

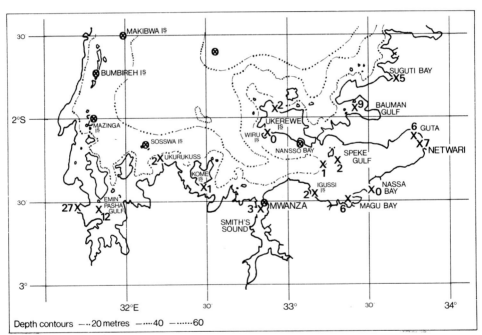

59 Southern half: ngege fishery. On this half chart are marked the catches of ngege per night, corresponding with papyrus shown on the previous figure. Positions where no fish of this species were taken are marked by a cross within a circle. Comparison of the two will show that the theory of their connection was borne out in twenty-one trials out of twenty-three. In the northern half of the lake, not shown, the results were nineteen out of twenty-three, still sufficient to let the theory stand. In each, there was one trial that was against the theory, showing that other factors do come in, but not very commonly. Evidently the prevailing wind, distribution of rainfall, scenery and papyrus and the ngege are all linked. Another factor is the distribution of lungfish which feed solely upon snails, which again feed on periphyton as ngege do.

60 Older ngege. In Fig. 56 it is seen that the Emin Pasha Gulf is set at the opposite corner of the lake from the Kavirondo Gulf, possibly therefore holding a separate stock of fish from that typified by Fig. 49. Comparison is shown up by this picture of what was undoubtedly an old ngege, one of many like it in that gulf. To us the appearance was so remarkable that we doubted at first whether the fish really were ngege, but they were. We were able to make a statement later published, in Graham (1943), which contains the essentials of most fishing theory.

Gulf	Fishing	No. of nets	Stock (per net)	Age	Yield (per night)
Kavirondo	Commercial	250	5	Young	1250
Emin Pasha	Primitive	0	25	Old	100

Evidently there was no cause to worry about the fishery, but, as a precaution, I recommended that there be a mesh regulation to prevent smaller fish from being taken commercially.

representatives of those fish that I had wished that I had seen in the early years in the Kavirondo Gulf. When the net fisheries had started, there was evidently not only the current year's new crop of fish but also an accumulation of old fish, and this accounted for those initial catches of 27 per net. Naturally, after a time, all the old fish were gone and only the new age-groups of young fish could be taken. Pictures of them show the differences (compare Figs. 49 and 60).

This is a situation with which we are familiar in Europe, when new fishing grounds open; at first there is an accumulation of old fish which are hardly worth marketing and it is only when these have been fished away after a year or two that there is a good, profitable fishery.

Although I did not realize it at the time, I had found the general solution to the overfishing problem, which I understood properly only some eight years later. The fishing problem is truly one of stock; the steady yield depends upon the average age of the fish in the stock and upon the intensity of the fishing effort—which, when it catches more reduces the stock left to catch. Hence there is an optimum over a period of time.

Although I did not realize that I had encountered the general problem, I was able to assure the governments that they would not lose their ngege fishery. However, I advised them that they should not, if they wished to retain the level of 5 to 7 per net, fish any more intensively, and should keep their mesh size at 5 inches. They did not all follow my advice about the mesh of the nets. Even in British Dependencies it was difficult to obtain agreement between three governments. Only the government of Kenya agreed and introduced mesh regulations, but these were widely disregarded. Everyone was satisfied, however, with the answer that the survey had provided, and four more surveys were carried out, by E. B. Worthington, my excellent assistant on Lake Victoria, and by his successors.

Retrospect

I confess that there was a personal incentive to drive: aged 30, I had not been inclined in the preparatory period to let anything stand in the way of what seemed to be a good opportunity to show what I could do. I considered that there were at that time no 'experts' in fisheries. What did exist were mere fish systematists and a few ecologists who knew only some of the things about particular areas they had studied. The whole subject was continually erupting with minor contentions and there was no acceptable principle or even theory available to apply to a new area.

That last consideration might well have deterred an entirely reasonable young scientist from touching the proposition. Distinguished men had turned it down before it was offered to me. Still another factor, therefore, was necessary to the decision to go ahead. This was my Micawber-like belief that something would turn up, what I have already called the unreasonable optimism that I have found so sustaining throughout life. As I have related, in the upshot something did turn up, in the Emin Pasha Gulf.

On the strength of that optimism, unreasonable at the time but shared, be it noted, by Professor Stanley Gardiner at Cambridge, I began to construct in my mind a research expedition. So as not to be peremptorily refused, I stated a time of seven months on the lake, which I thought to be as far as I could stretch the sponsor's probable initial idea of perhaps a seven weeks' tour, and the longest I could saddle my colleagues at Lowestoft with my share of the sea time of research vessels, which, when once acquired, have to be kept busy, or they lose quality. In specifying seven months, I had taken some control of the scope of the expedition in time, but I had no room to manoeuvre about its coverage in space. I had learnt in the sea that one must study as much of the fish population as was liable to turn up in the area of research. The charts of the Victoria Nyanza showed no natural barriers cutting the lake into portions, nothing that would stop a slow interchange of local fish stocks. So, the range of space to be investigated was the whole lake. A small outfit such as I had seen worked from houseboats and a barge on Lough Derg in Ireland by Southern and Gardiner, would not serve. Their work taught me one thing, on which my fixed station of Sukuri Island was based, and much of my later work as well: namely the considerable progress that can be made by intensive study in one spot.

The determinant
In the preparatory stage in England, I had, knowing nothing better, planned to attempt everything we were doing from Lowestoft, adapting it to the conditions of Lake Victoria. At the time at Lowestoft, Dr. J. N. Carruthers, who was later at the National Institute of Oceanography, was doing well with wind studies and drift bottles; so I planned the same for Lake Victoria, and hit a bull's-eye, as it turned out. The south-east trade proved to be what we have in this book been calling the determinant. It is the key, even to the scenery, as well as to the ngege distribution, as may be seen in our chart Fig. 56, based largely on observations by Indian piermasters of the Kenya and Uganda Railway Marine.

Scientific research
In later years at Lowestoft, I had a spell of upbraiding the staff for trying to approach discoveries directly. In return, I was criticized, for that was their wont, for the performance on the Victoria Nyanza, as related here in the lecture, in that I had proceeded crab-fashion: working on one problem, namely the topographical distribution of ngege, in order to solve another, age and catches. The procedure did not seem logical, nor is it. Logic cannot cater for the totally unexpected. I would make a principle of that, and say that if a scientist can see where he is going, it is hardly research that he is doing—more like development. In the history of discoveries, a great many of them appear to be accidental. We should foster something like unreasoning optimism if we are to make exciting progress, which is the discovery of the totally unexpected.

However, an immediate comment is that the unexpected discovery is not purely accidental, any more than it can be purely deliberate. This is illustrated by the examples in which the discoverer has not recognized his luck. To notice significance, he must have worked hard in the chosen field where the discovery appears, and must have thought all round his subject. To be 'happy go lucky' or to sit back and wait for an accident will usually be barren; yet the discovery comes unsought. It is unproductive to search step by foreseeable step, because, in a multi-factorial world it is usual for the step by step to omit some important factor.

Evidently, discovery is a mixed process: it may fairly be called an art. However, a critic may continue the attack, and ask why it is necessary to take so much trouble in testing an hypothesis, which when confirmed is found after all to be not decisive. He might argue plausibly that the same luck might come into a programme of unco-ordinated observations, of general survey, with which we began in out first circumnavigation of the lake.

There is no short refutation of that argument. I can say only that it has not seemed to work that way in my own experience. Unco-ordinated observations put me in a muddle.

At the earliest stage, however, there is a world of facts to be gathered. One needs to make as it were a small encyclopaedia and I would again advise ecologists not to neglect this preliminary stage.

Looking through the section on general ecology in the published report on the Victoria Nyanza, one finds notes on chemical analysis showing high sodium content in the lake and a high pH in the open waters, up to 9·0, but related to a low alkalinity by analysis. That is typical of the interesting but not immediately useful facts that one collects in general gathering. It would be tedious and unprofitable to go through unco-ordinated facts here but they were quite exciting to collect. For instance, at Sta. 29, when trammel nets were fished on the bottom in 216 feet of water and gill-nets were used on the surface, stomachs of the fish called *Synodontis* caught in the surface nets contained emerging adult gnats, etc., as well as the larvae and pupae, while those from the bottom of the lake did not contain the adults. The fish were apparently rising through the water following the gnats as they metamorphosed.

The principal food of this fish seemed therefore to be the larvae and other stages of the insects. The commonest of these are the 'lake-flies' (*Chironomids*, i.e. gnats, chiefly) which are present in great numbers and produce an appearance of water-spouts on the open lake. When they drift ashore on to a township there is such a mess over and in the houses that the effect is one of disaster.

That example of unco-ordinated observations is thus interesting enough, and perhaps justifies the place given to *Synodontis* as a principal converter of the insects living in the ooze, and so as a major factor in the nourishment cycle of the lake. But the information is not positive. I doubt if the number and arrangement of observations is adequate. The provisional conclusion is not better than a well-founded surmise: especially when compared with the conclusions from observations arranged to test the hypothesis of ngege distribution.

I am inclined to think that unco-ordinated observations have a proper role as a basis for future research, and no more than that.

Awkward facts
Whether by co-ordinated or unco-ordinated observation, one needs must gather as much information as practicable about the ecosystem, lest, unbeknown to the investigator, there lurks in some niche of it an awkward fact, such as the one George Morrow drew (Fig. 61) in *Quoth the Raven.* In science, awkward facts have important roles: when confronting current beliefs, they open the door to new *paradigms,* as the word is, metaphorically one might say that they form new view points to undiscovered country. They have also the role of monitoring new discoveries.

In the argument, man is taken to be the top predator, but we may consider other possible candidates. We did not find Nile 'perch', *Lates,* in the lake nor at the Ripon Falls, which form a barrier, possibly secondary to that of the Murchison Falls, but which are more often under observation, and local information agreed with ours from line fishing: there were no Nile perch. Worthington proceeding to Lake Albert, took them there with our Lake Victoria gear. We could be satisfied that there was no awkward fact lurking there in the form of Nile perch. Had there been Nile perch, the solution in terms of man and ngege would hardly be adequate. However, in science the position is never secure. We saw a white-brested fish eagle *Halietus vocifer* with a large *Tilapia* in its talons. Many eagles could form a factor; but we saw this predator once only; and other fish-eating birds were harmless to ngege, or neutral, if we judge by their stomach contents.

So, there is no end to the possibilities of an awkward fact, but taking numerous unco-ordinated observations makes its undetected existence less likely. That is all that the ecologist can do.

In practice, there was a sad sequel to my efforts. An inspector was appointed in Kenya to enforce the 5-inch minimum mesh that

61 Escape of an awkward fact from the Press Censor's office. George Morrow's cartoon is not a digression but a reminder that science often advances through the escape of an awkward fact. Our awkward fact was the manifest advanced age of the ngege in Emin Pasha Gulf. It should have been a warning against any regulation, but the thirty-year-old investigator bowed to the then fashionable doctrine and recommended a mesh regulation. It was not enforced and the mesh in use fell to four inches instead of five. The yield rose still further, and the lower catch per net did not hurt the operators, because non-rotting nylon nets came into use and they could fish more nets. Unfortunately, a new crime had been invented, serving perhaps as a warning needed by ecologists of the future. (By permission of Messrs. Methuen & Co. Ltd.)

I had recommended. Who could know more of these matters than my fishing assistant on the survey, whom I had liked well? The next I heard was that he had become very fat in his new post; that the customary mesh was now 4 inches and the catch near Kisumu 1 per net. After a few more years I heard that my former man had landed in gaol, and had become much thinner again. The low catch of only 1 per net could now be tolerated because nylon nets outlasted the linen ones and so reduced running expenses: all this is hearsay.

The choice of the next lecture was now plain. Aquatic ecology had shown changes in sizes and numbers of plaice; woodland ecology had mentioned numbers in the Elton pyramid, which we had taken further in the study of predators and prey. Now comparisons of ngege populations had brought in their relative numbers.

Moreover, the acceptability of the Victoria Nyanza lecture allowed me to see that lectures about my own work would not, as I had feared, be taking advantage of defenceless students. They liked it.

It therefore seemed natural that the next lecture should be on estimation of numbers in wild populations and related problems that had formed a large part of my working life.

Population numbers   11

Morand and Laplace

'How to estimate from the birthrate the population of a vast empire.' That is a free translation of the news that broke on the Academy in Paris in 1771 in a memoir from a magistrate named Morand. The interest was continued in that quarter for about twenty years.

Morand's work was taken up by Laplace, a prominent academician, who had taught mathematics at the School of Artillery, where Napoleon the First was one of his pupils. Laplace took care to send complimentary copies of his books to the great man and received courteous replies, one of which said that Napoleon did not understand it but was nevertheless very glad to see it. Arago, a biographer of Laplace, tells us very little about the personality of the man, except that he was rather fond of appropriating other people's mathematical discoveries, that he concealed his humble origin and in spite of the troublous times was honoured by being created a Marquis.

Morand's method, which was not wholly new, took advantage of the fact that from time immemorial births have been accurately recorded, in France as elsewhere, and he first estimated the ratio of births to total population from samples spread over France, spaced so as to compensate for any effect of variation in climate and circumstances. He was careful to select communes in which the mayors were reliable men who would give him accurate figures of the ratio of births to population in their districts. He found that the ratio, although it varied by half a point or so, could in all districts outside Paris and Versailles be represented by the factor 26. In the metropolis, there were more people who did not become parents, such as monks and nuns, and these were listed in his table and brought the metropolitan ratio nearer to 31. That reference to the clergy may have given rise to the story that Laplace's method (so-called) was to ascertain the ratio of monks to laymen in the Paris streets, and then, knowing the number of monks in monasteries, multiply it to get the total number of people in Paris. However, this story seems to be mythical. The staff at the Chetham's library most helpfully found Morand's papers for me but I could trace no record of the counting of tonsured heads.

Hensen

Two men in the late 1880s were making assessments of populations of fish, and it is difficult to say which should have precedence. Perhaps, strictly speaking, Victor Hensen of Germany should, because in 1887 he outlined a programme of quantitative plankton work. In any case, his is the method to mention first because it resembles that of Morand.

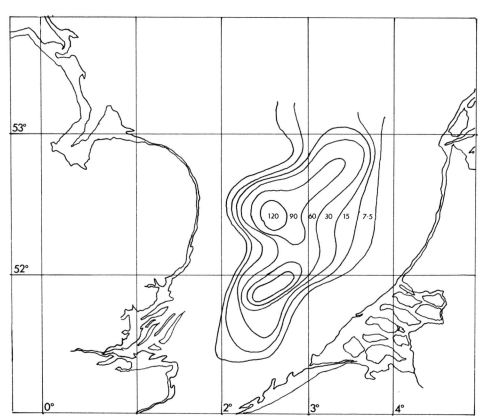

62 Census by eggs. If the contours close, as they nearly do here in Wollaston's diagram of Hensen's earlier idea, biomass is estimated by spot counts of planktonic stages of the animal, namely the eggs of plaice under a square metre, integrated for the total area. This is a sound idea and works. It provided valuable confirmation of the order of magnitude of the number of plaice in the fishable population in the North Sea—about 300 millions. Those plotted here would constitute the spawn of about half of the females, the outermost contour being of 7–8 eggs under a square metre of surface, and the two innermost contours representing 90, with a peak 120 after Graham (1956).

His method is a remarkable one and perhaps not sufficiently well known, but in a way it was the same as Morand's because he also tackled the problem by taking a stage in the life history that is easy to count In Hensen's case it was the eggs of the fish. It was Apstein and Buchanan Wollaston who made Hensen's idea work, but it was surely his idea (Fig. 62).

Most sea fishes have eggs freely floating in the water and by a filtering of a known volume of water one can count the eggs in a vertical haul, making adjustments for filtration efficiency, and then the numbers under one square meter of surface can be put on a chart and contours drawn and the total number of eggs in that part of the sea can be obtained by graphical integration, just as one can obtain the volume of a mountain from a contour map of its heights. In that way one has an estimate of the total number of individuals at the egg stage and it is only a matter of having the right information on the average number of eggs of the female fish—this has been done particularly well for plaice—to have an estimate of the mature females. Then, by going to the fish market, one can obtain ratios of mature females to the rest of the fishable population.

Without going into great detail, we may quote a figure of fifty million plaice spawning in the North Sea. This figure was obtained through the work of Hensen's successors in this field, especially Buchanan Wollaston and A. C. Simpson.

The excellence of Hensen's method lies in the comparative uniformity of plankton, not that it is perfectly uniform, but usually determinations of organisms under a square metre of sea surface do contour satisfactorily when they are put on charts.

The corresponding egg method has been tried for insects but the terrestrial habitats, such as trees, bushes and herbs, are more difficult to sample than is sea water, because less uniform, but it is only fair to mention that counts even of fish eggs in sea water show some patchiness, but not enough, evidently, to invalidate the method: otherwise the figures would not contour.

Petersen

The other pioneer working at about the same time was C. G. J. Petersen who published in 1894, and was probably the first of all naturalists to attach marks or tags to wild animals in order to obtain population characteristics by recapture.

A fishery naturalist, G. T. Atkinson, about ten years my senior, told me of the usefulness of the early estimates of population in convincing fishermen and other people of the necessity for regulation, even in the great areas of the sea. It was necessary to have some sharp counter to the view of T. H. Huxley who in 1883 had advised fishermen to fish where they liked, when they liked, and how they liked, without let, hindrance or concern, because the numbers of fish were so vast. Atkinson said that it always made a planning marking experiments, it was necessary usually to estimate for expenditure on rewards for returned fish on the basis of 40 per cent of those liberated. Such a striking fact made people listen to the rest of what he had to say. Thus Atkinson was following Petersen's lead.

So, too, doubtless unknowingly, was F. C. Lincoln, who in 1930 published his work on putting bands on ducks, originating what many ecologists call the 'Lincoln index'.

Let us suppose that Lincoln put bands on 100 ducks, then after the ducks had distributed themselves throughout the area he started to shoot them and counted the percentage of bands on those he shot. Let us suppose that it was 10 per cent. At once we can see, intuitively and without any algebraic statement, that 1000 would be a fair estimate for the number of ducks in the area. Lincoln's is also the so-called Laplace method, which I believe to be mythical in which the men with tonsures would be the marked or banded specimens; their number could be ascertained by going round the monasteries, while determinations of their percentage in the population could be made in the streets.

Direct methods

There are some other methods of estimating the total population. I once used a direct method when I wanted to discover how many gulls were attacking the *Euphausiids* and other plankton that were coming up in a vortex in the entrance to Pasamaquoddy Bay on the borders of Canada and the United States. I took photographs of the gulls in the air and thus was able to count the birds. It came to a very considerable number, but of course would be an underestimate because some would at the moment be actually down on the water and would not show on my photographs of the sky, and there might be other members of the same population who were not at that time on their feeding ground.

There is also the direct counting of animals as they go by a point on migration. That has been done visually for caribou, which are wild reindeer, and for salmon passing a weir, but a refinement is to record them electronically, which our Department of Biology is doing, using the device installed by the River Authority on the River Lune.

A student at Salford, A. J. Taylor, placed boards at various points on the bed of the local disused canal. This made the roach easily visible and he then multiplied the average number of fish that he saw on each board, on first reaching it, by the ratio of the area of his boards to the area of the canal. Of course this method was subject to variances of behaviour and not very accurate, but in Chapter 5 we learned that it gave a surprisingly large number of roach, namely 5000 of 8–12 inches long in 380 yards of canal.

On land there is a similar method called the quadrat method, in which one puts down frames of exactly one metre square and so counts the organisms per square metre and then multiplies up for the whole area, just as for the canal.

Southern
A considerable volume of work was carried out following Petersen's lead in fisheries and there has also been a great deal of work since Lincoln's paper on the ducks.

A good example of the work on mammals can be obtained from Southern's description of how he and co-workers deal with estimating populations of small mammals. The plan is to mark once a month and to recapture a month later, over a period of perhaps nine months. Those who would repeat it must read the author's own account, but two points will show how powerful is the technique. The awkward questions of emigration and immigration and of mortality and increase of numbers by reproduction are encompassed. Southern points out that if for the first two or three months of liberty the rate of return fluctuates around 10 per cent, and then rather suddenly in subsequent months takes to fluctuating round 5 per cent, one can suspect that the number in the population has been very greatly increased, and still obtain useful information from this population, although it has in fact doubled at the breeding season. Southern points out that it is thus, from the successive figures of returns from each marking, that one detects immigration or increase of the population by reproduction. Emigration or very heavy mortality is obtained from changes in the percentage return in successive months. For example, if marking 100 mice in August gave a 10 per cent return in September, but marking 100 mice in September gave a 20 per cent return in October, one would say that in September the marked mice have formed a greater percentage of the total population, which must therefore have fallen.

Improvements

My colleague at Salford, Michael J. Parr, and some students, carried out an experiment at Dale Fort (1965). He worked up the returns of *Ischnura elegans* the damselfly, which is the slender beautiful type of dragonfly. Parr demonstrates several different ways of working up the results, all variations on the 'Lincoln Index', and finds the number flying about on any particular day.

All the methods described so far have used the numbers of marked specimens liberated as the start of calculations. But that number becomes invalid very soon, because marks are lost, specimens may not be returned, they may have been fatally damaged by marking, or upset and frightened away from their usual haunts. There may have been emigration or fresh births, or variation in the natural mortality rate. If a mark makes them more conspicuous to predators, it would affect the estimate of the marked population. Much research has gone to mitigate effects of some of these errors; but one modern method, Gulland's (1955), has greatly reduced them and another, Manly and Parr's (1968) still more so.

J. A. Gulland worked in two steps. First, he took into consideration only returns that came in very quickly after the liberation, so that any errors operating during the time at liberty were very much reduced. Taking only the small area in which the plaice could have distributed themselves in a short time, he determined a co-efficient for the percentage returned and the deployment of fishing power (horse power and number of standard vessels, etc.): that was Stage 1. Proceeding with Stage 2, he had statistics of fishing power over the whole North Sea, which was a much larger area. Using the co-efficient from Stage 1, he was thus able to determine the rate of fishing over his whole area and from that the total population, by applying it to the whole number caught.

B. F. J. Manly and Michael Parr have introduced a method that goes further. In many cases it could eliminate all the effects mentioned above. It also works in two stages. It uses an estimate of the proportion of the population of Burnet moths captured between July 19–22 in Pembrokeshire on a sunny weekend. The intention was to work only in the sunshine, which made all the moths active and vulnerable. Each day's capture was marked with coloured dope on the wings—Friday's green, Saturday's white, Sunday's blue and Monday's orange. On the following Wednesday, to round off the experiment, they hunted and caught moths but did not mark any of them, only recording the ones marked previously.

Let us consider the data for the Saturday's work. After the experiment is all over, the calculation of the proportion of capture to the population on Saturday will be made only on those of the previous Friday's markings which the records show to have survived to Sunday or later: the survivors after Saturday. This is precisely known; it numbered twenty-six moths. Of the twenty-six, fifteen were recaptured on Saturday, which makes the estimate for recapture rate on Saturday 15/26. The total number caught on Saturday was fifty-two, and it is therefore clear that the population on Saturday was almost exactly ninety. Three such estimates come from the five-days' work, they are ninety, ninety-six and ninety-three, leaving out decimal points of a moth.

Manly and Parr's method seems so good, as well as so simple, that it must be far and away the best yet, whenever circumstances allow the experiments to be arranged in that way.

It may still seem remarkable that men are able to assess the number in wild animal populations, so that a man can say how many fish there are in the sea to within a million or two, how many mice there are in a wood to within a few hundreds, and even how many tsetse flies would have to be destroyed to eliminate them altogether from an area of shady shoreline. Nevertheless, wonder should not amount to disbelief: once the methods are understood, the estimates have an air of authenticity.

Petersen's intuition of the optimum catch

Having thus demonstrated the power of ecologists to estimate numbers in wild populations, we are in a position to return to the theme we have been pursuing through the last three chapters, namely predators and prey. We will do so by instancing the well-worked example of men and fish. Indeed J. Z. Young has done this so well in *The Life of Vertebrates*, that I would take his chapter there as readable by many biologists and others with some simple skill in algebra, and attempt to widen the circle of understanding among those less skilled, by using a little history and by explanations without algebra.

The fact that history provided modern workers with the requisite data is due to the intuition of our predecessors, among whom I would again mention Petersen, who in 1894, wrote: 'It cannot well be doubted that the same area of sea will be able to give a quantitatively greater profit as a constancy when we suffer the stock of fish to be as fully developed as in the years before the too-eager fishing commenced and then took exactly as much as the stock could reproduce by new growth.'

Thus was the central problem correctly stated. Good as that statement still is, men demand quantities. An international meeting was convened in 1898 and a permanent research council established in 1902. Under its enthusiastic care all the North Sea fishing countries collected data on all the important species of fish, with a greater or less degree of success, and the council's activities were then extended beyond the North Sea to other areas fished by European nations. Haeckel's ecology, autecology, was pursued in every possible direction, and synecology when possible. Nobody, however, knew how to put the data together so as to produce quantitative estimates and I fear that the majority of scientists in the meanwhile, and some even at the present time, overlooked the essence of Petersen's statement that it is in terms of profit. E. S. Russell who inducted me into the subject, realized that the index we used of biomass, the catch per day's absence of a standard trawler, was also closely related to profit. Dr. Russell devised a very simple, almost non-mathematical equation. He said that we should consider the case where a stock is in equilibrium with the fishing power and does not change its average weight. In that case he said that $A+G$ must equal $C+M$, all expressed in

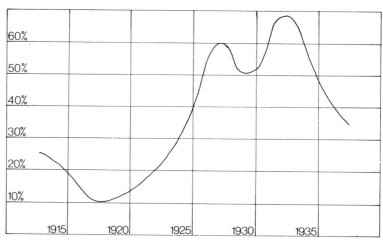

63 Total mortality rate: Netherlands soles. Crude figures, if collected regularly, can provide definite information: this information is on change in total mortality rate. It is made from the percentage classed as 'small' in market returns of soles. These then are small soles from Holland. The late Sir Darcy Thompson every year commented on the corresponding figures from all species and countries at his statistical committee of the *Conseil International pour l'Exploration de la mer* at Copenhagen, from 1902 onwards. They invariably showed this type of change—an increasing percentage of 'small'.

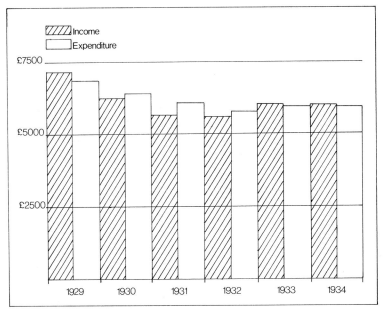

64 Overfishing. The crunch came in the North Sea and neighbouring waters in the 1930s. The diagram is based on figures from the White Fish Commission, giving the average annual income and expenditure of Grimsby trawlers fishing near and middle waters, 1929–34. When this evidence is accompanied by that on increased fishing effort and fall in catches, there is clearly too much fishing.

weight. A was to be the weight of the recruits entering the fishable stock at a given mesh of net—new young fish; G was to be their total of growth during the period of time; the catch, C, was to be the weight caught and M the weight destroyed by natural mortality. Then he turned the equation round and said that in equilibrium

$C = A + G - M.$

That is to say the predator could take a nett increase of weight in the population and no more, if the situation was to continue unchanged, and the natural increase of weight would be compounded of the nett effect of processes changing the weight.

He next asked himself how the catch could be increased and pointed out that if recruits were allowed to become heavier before being caught, through an increased mesh size, there would be some increase in the weight of the catch, provided that the heavier weight of fish in the area did not sufficiently reduce G, the growth increment, or increase M, the loss by natural mortality. But Russell saw no way to estimate these changes due to changed density, and stopped at that point.

A special point of Russell's work, which had been foreshadowed in Petersen's dictum 'as a constancy', was of equilibrium, which could in practice only be achieved in a regulated fishery. This is anathema in certain circles. But the assumption since the very early days of the agitation of the smacksmen has been that we need regulated fishing and that the gains would outweigh the various frustrations and disadvantages involved. Otherwise we come to a melancholy unintentional equilibrium, which is illustrated in Fig. 64 shown here for income and expenditure of a Grimsby trawler fishing in near and middle waters in 1929–34, gradually approaching an equilibrium fluctuating about a profit of nil.

Although equilibrium, as an average over the years, is a real feature of many commercial fisheries, there are also real fluctuations, some of which Young has shown in his chapter and which Simpson has included in *Sea Fisheries*. Their existence does not invalidate advice based on the analysis of equilibria, which still shows approximately the best way to utilize a fluctuating fishery.

When Russell gave up his line of enquiry I was disappointed, but did not see any way forward. It was therefore very encouraging when a couple of years later two papers from overseas came to my desk for review. One was by W. F. Thompson (1936) who had persuaded the Pacific halibut fishermen of Canada and the United States that they were fishing the stock too hard for the best profit, and to agree by a convention to limit the catch to a certain total quantity. He then found that the fishermen were reducing the effort to take that quantity every year. His theoretical basis was an arithmetical one; that is to say, he gave a good deal of reasonable arithmetic to illustrate Petersen's original insight.

The second paper was from Hjort, Jahn and Ottestadt in Norway, *The Optimum Catch* and was about stocks of whales. It did not assume rates of increase or mortality constant but followed a reasonable theory as to how they would vary with density of stock.

They considered a stock of whales entering a new area or recovering from near extermination, and assumed that its rate of natural increase would be proportional to the difference between the weight of the stock and the upper limit that the area would carry. Theirs is called the autocatalytic or sigmoid theory and it has recently been revived by Gulland and used to calculate the best levels of fishing for the whale stocks under exploitation in these modern times. I illustrate it here by one of my own versions (Fig. 65).

It seemed to me that these two sets of authors, whose theories look extremely different, must be giving two aspects of the truth, and Russell's equation must fit in, so that if one could integrate the three papers on theory, one could probably apply it to the parlous state that the North Sea fisheries were experiencing at that time. We had no doubt of that, but it was only later, from the work of the White Fish Commission that we could document the working of Petersen's conception, as in Fig. 64.

The application of the theory of fishing to real data is not lengthy, as such things go, but an imaginary example can be much shorter, and thereby show up all the better the essence of it.

In far Rainbow Land there is a township called Troutville, because it is situated on the shore of a lake in which people fish for trout as their main recreation. There must be 400 fishing rods there, and they catch on the average 46 kilogrammes of trout per week for every 10 rods. By reading age from scales, they know that their catch is constituted on the average as shown in lines 1–4

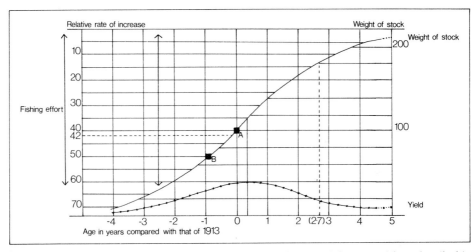

65 The optimum catch. In a way this is a sequel to the history of the ngege fishery described in Chapter 10, but the story is by no means finished yet. As the ngege fishery developed, yielding more in total, the average age of the fish fell, the weight of the stock dropped, and consequently so did the profit represented by the catch per net. In this diagram, in which sixth-formers will recognise both logarithmic growth and differential calculus, that history is formalized. The central theme is that the stock has a potential rate of natural increase, growth and reproduction less mortality, at each age, which natural interest can be taken without harm. The best state is somewhere at a middle of intensity of exploitation. This has been plausibly advanced for several natural resources—yeasts, people, whales, timber, and forest game. It is probably nearly universal; but how to determine the precise optimum point (not always the 'maximum sustainable yield') and how to share that optimum catch are two essential problems not yet determined theoretically. Nevertheless in practice, it may be possible: it was said that Rasmussen organized something among Eskimos of Thule whose staple walrus was threatened by modern rifles. The main lesson of this diagram is that, to take full advantage of the natural potentiality of the living world, stock must be allowed to build up, whereas it can all too easily be cashed. It therefore needs a moral order in society, to outlaw greed.

of the Table (below). And so the fishery continues as the years go by: every year, 6 recruits provide, when growth and mortality have played their parts, the 46 kilogrammes. The total mortality includes that inflicted by the 400 rods. One kilogramme of recruits provides 46 kilogrammes of fish, and what is more important in the minds of the city fathers, many hours of innocent occupation for the citizens.

The people of Troutville have no anxiety about the supply of recruits, because the trout breed in the undisturbed streams running into the lake, and there never seems to be any shortage in supply of the young.

To a visitor who thought four and a half kilogrammes of trout rather a meagre reward for a week's fishing, his hosts replied that he should take a week twenty miles away at a holiday hotel, which controls a very similar lake, the same size and a similar drainage basin, with the same species of trout in it. There he took nearly seven and a half kilogrammes as in lines 4 and 5 of the table. Possibly this might be due to a natural difference between the lakes; but the scientists had marked many fish, and so had investigated the mortality by fishing and its ratio to the total mortality, and they found it perfectly reasonable to explain the difference as due to greater survival from year to year stemming from the hotel's deployment of only 200 rods against the town's 400.

Thus too it was for the staple North Sea species. It would be tedious to chronicle the whole matter here, when all the elements

	Age in years								Total
	1	2	3	4	5	6	7	8	
1. Number (A)	6	4	3	3	2	1	1	0	20
2. Av. wt. (kg.) (A+B)	0.17	1	2	3	5	7	9	10	
3. Wt. of stack (kgs.) (A)	1	4	6	9	10	7	9	0	46
4. Number (B)	6	5	4	4	3	2	1	1	26
5. Wt. of stock (kg.) (B)	5	5	8	12	15	14	9	10	47

A simplified illustration of the theory of fishing 1935. Lake A is supposed to be furnished with 400 rods. Lake B with 200. For each lake the history of 6 recruit fish is traced through their 'life-tables' over eight years, lines 1 and 4, as sampled by, shall we say, 10 rods per week. It noted that the number falls off more quickly in A than in B and we are assured that marking experiments have shown that this greater mortality may be reasonably ascribed to the greater number of rods. The ratios of the total annual catch are as 92/74, but in their circumstances each state is sensibly exploited: recreation is the main objective; A provides evening and other short refreshment to townsmen; B requires exciting sport for holiday-makers.

the general student needs to know have intruded themselves into the imaginary example at Troutville. The real data for North Sea trawl fisheries are in my paper of 1935, in the technical literature.

Even the true nature of rational fishing is to be found at Troutville—namely a matter of macro-ecology, in that it is social. The town gets its recreation daily; the holiday area is less crowded, and provides better diversion for shorter spells. Evidently both are rational fishing. I once addressed some Spanish fishermen, who work appallingly hard, as most fishermen do, thus: 'Sleep fishermen, sleep. While you are sleeping the fish will be growing.' That is the implication of the theory of fishing: the economy of fishing effort both in men and in materials, wear and tear of ships and gear and expenditure of fuel oil would begin to operate at once. This illustrates the argument that the modern theory promises towards the profit equation greater economy of effort than increase of catch. Thus it catches up with Petersen's statement.

The Table uses unaltered rates of growth and mortality in spite of greater biomass, and it was necessary in the paper of 1935 to show that alteration could hardly be expected to invalidate the sign of the change of yield on reducing fishing—as well as to back the conclusions by other data. At the end of Chapter 9, we saw that there is a suspicion that in some populations the greater biomass of fish—74 compared with 46—would attract more predators, and that the fisherman would not obtain from his effort the same proportion as he did before. That is certainly to be expected. A reduction of fishing effort, owing to reduced profits, has sometimes caused fishermen of other countries to obtain a slightly better profit than they were doing, and so to increase their fishing effort, rendering null the total benefit of 'rationalization' of those in the first nation.

The calculations of 1934, represented here by the imaginary example, coupled with the macro-ecology that was included in the same paper (1935), convinced Russell, who was in consultation with Professor Raymond Pearl, and went over to lecture to his students in America. The result was Russell's book *The Overfishing Problem*. After all these years of work by our predecessors, it was possible to advise governments to arrange for an international control of the rate of fishing, on top of the agreement that had already been reached that there should be mesh regulations, specifying that the ones most commonly used should go no smaller.

With the help of scientific colleagues and friends, especially those in the Challenger Society of Oceanography, of whom I will mention Alan Gardiner and Stanley Kemp, the British government was persuaded and the policy agreed by all parties, so that the administration was ready to put it into practice as soon as possible after the cessation of the Second World War. Unfortunately, the international world refused, for various reasons which have not much to do with ecology. Let us merely say that as the flush of victory faded, so did enthusiasm for planning a better world. We achieved a Gentleman's Agreement but it was dropped. This was very disappointing, especially when one remembered the words of Sir John Murray, an oceanographer of world-wide distinction. He attended the inaugural meeting of the International Council for Exploration of the Sea as the representative of the British Government. In his speech he said that he foresaw the end of economic nationalism and envisaged a time when, as the best thinkers of his generation agreed, nations would not be divided by frontiers, and all would co-operate for the common good. Two generations of naturalists had worked hard to solve the necessary synecology and it had not been used. All we could collectively do was to resolve to try again later.

In the meanwhile we considered what to do, helped by P. M. S. Blackett and H. R. Hulme. The upshot was the immortal algebra of Beverton and Holt. Sidney Holt carried the theory to Rome when he was transferred to the service of the F.A.O. He was later joined by John Gulland who had pursued the same line at Lowestoft, and who was later to restate the theory of fishing as applied to whales.

Beverton and Holt's book on the subject (1956) took eight years to prepare and write. They had mastered the subject in two years but the remaining six were taken up in searching world literature for examples of different situations which they had envisaged. As a result their theory can be used to find a solution to data of any problem of fishery regulation in the world. It will, however, have different emphasis where reproduction rate is important.

Looking at Beverton and Holt's work in retrospect, I would pick out their conception of eumetric fishing as being outstanding: that there is a best mesh for every rate of fishing and a best rate of fishing for every size of mesh.

A major achievement of Beverton and Holt was the inclusion of Bertalanffy's physiological equation for growth. It was a technical necessity to have some equation for growth and Bertalanffy's seems particularly good in representing physiological reality, in anabolism and katabolism.

I would claim that the particular value of the English fishery work is that it has thrown up various confirmations from independent lines of evidence.

Marking experiments gave a rate of fishing of 0·69 for the years before 1914. The definition of the rate is that fishing was taking 69 per cent of the average stock—while it is being continually replenished—the model being like a reservoir. The wartime cessation of fishing (1939–45) allowed Beverton and Holt to estimate the natural mortality for certain year-classes which survived through the period of non-fishing. The best estimate they had was 0·1, which, if added to the previous rate, gives a total mortality rate of 0·79. It is true that the addition is made by combining data not from quite the same period. For the total period 1929–38, again a somewhat different one—the estimate of total mortality from the life-table as in those of Troutville above—was very close: 0·83. To my mind the approximation of the estimates from these data is gratifying, especially knowing intimately all the difficulties of collecting them. In fact, the later figure would be expected to be somewhat higher than the other, because of the greater efficiency of the fishing fleet which in the later period consisted of steam and motor craft instead of steam and sail. In addition, the trawls had improved in catching power.

Then we have the evidence of the egg census giving us a crude 385 millions as the stock of fishable plaice, as against the life-table plus wartime calculations of 337 millions, only about $12\frac{1}{2}$ per cent less.

After the Second World War an international meeting was held to report on the resulting changes in fish stocks. In all areas the changes were in accordance with expectations, and the expectation had been derived from investigation of the results of the First World War, and a number of further studies. None was discrepant.

Much later when a number of older English trawlers gave up fishing for plaice in the southern North Sea, the stock increased by the amount that Beverton and Holt's calculations would have predicted for the corresponding reduction in rate of fishing.

These scraps of confirmation that we have just given seem to my mind to place fishery science or fisheries economics on a level above that of any theoretical work that has no confirmation.

We may look back over the history of fisheries science and wonder if the limitation of the technical work was realized. It is clear to see now that, however valid and excellent the synecology may be, to carry conviction it must be accompanied by macro-ecology, such as historical evidence, or the evidence of a deliberate experiment. When experiments involve human beings they are usually held unjustifiable on the grounds of correct behaviour of man to man; nevertheless, when in desperation a trial is made, as on the Pacific halibut, it does put the whole subject up to a higher level of acceptability and confidence. Thompson's work on the halibut (1934 and 1936), reaches a position of confidence that no amount of clever arguing can disturb. Thus halibut regulations were started and as a sequel the profitability of the industry improved over several years. There is no getting away from that. Nor I would say, is there any doubt of the lesson of the history of whaling.

We have been dealing with common food fishes whose rate of recruitment seems to be almost independent of the number of adults, just as you may have innumerable windborne seeds from one mother tree left standing when a forest has been felled. The history of whaling, however, is different, depending upon the rate of replacement by numbers. It is a particularly bitter story, beginning in the last century for northern kinds, then moving to the Antarctic. Species by species the larger whales became too scarce to give profitable hunting, proceeding from the right whale to the blue whale and down to the smaller rorquals (see Fig. 66). There has been an international commission but no effective agreement, no agreement with a cutting edge to it. Instead there has been great international jealousy and infractions of what poor agreements were supposed to exist. Immediate profit has all along ruled, resulting in no profit in the end. In the eighteenth and nineteenth

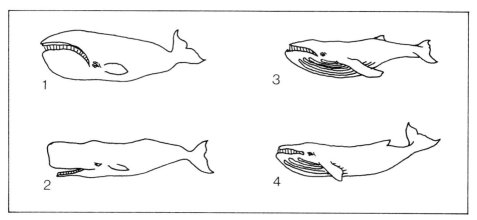

66 **Whales.** There had been a great deal of vain discussion about the depletion of whale stocks, but macro-ecology, as history, took charge by the successive exhaustion of stocks, species by species, from the Right whale (1) to the less productive ones: Sperm (2), Humpback (3) and a large kind of Rorqual (4), of which lesser kinds followed.

centuries sperm whales were reduced in that way until it ceased to be profitable to hunt them, and then the stock slowly recovered and a small-boat fishery survives from the Azores. That is perhaps hopeful for other stocks. On the other hand, on the grounds of its cruelty, one hopes that whaling will die out altogether unless electrocution or some other humane method of killing becomes practicable. Hardy's account of the killing in his book *Great Waters* is an extremely moving one, and one cannot help wondering whether modern drugs could be used on the harpoon, either on the first one or on a second.

Fish husbandry
It is not surprising that in view of the political and technical difficulties of international co-operation, in regulating fisheries, men's minds have turned to the problem of farming flatfish within an area of sea exploited by one nation, if such can be found. Cultivation of at least mussels and oysters to solve eutrophism, is a possibility mentioned at the end of Chapter 5.

Leaving fish-farming aside, fisheries, including those for whales, provide a unique case where studies have resulted not in merely understanding the macro-ecology of the fish stock but also the relation to human ecology. In fisheries, man can, if he so wishes, collectively harmonize his activities with the natural increase of the fish for the great benefit of everybody concerned, on a world scale and irrespective of the barriers that separate nations.

Technically, it is quite easy to see what should be done. Each whale gunner should have so many shots in his locker, when he leaves for the Antarctic, and each fishing vessel so many days horsepower's absence allowed during the year, which would be endorsed on his licence. Such crude ways would be effective enough for what is after all a pretty crude problem. Taking into account all the natural factors of variation, such enforcement should not be impossible in some countries, in others there might have to be close seasons, total fishing power held, or some equivalent way of rationing the fishing effort. Human knowledge is available for this great co-operation between man and his environment. No irreversible damage has been done as yet to stocks of fish. The sea is nearly unharmed and fish stocks can be put right by mere resting. Even the reckless pollution mentioned in Chapter 5 is absorbed, with some important exceptions. On land, however, the damage will not be reversed by nature's agency alone, except exceedingly slowly, and the situation calls urgently for constructive human action, as I shall try to explain in the remaining chapters.

12 Erosion/I/general notes

Magnitude

'The canal was full of ships,' said my grandson. 'We saw them from the motorway.' His route had brought him over the long high level bridge.

'Not really *quite* full of ships?'

'Yes, quite full.'

'A little ships' regatta perhaps.'

'No, very big ships.'

'For the canal that would mean five thousand tons.'

'Yes, head to tail.'

'Stem to stern you mean, but pilots would not advise captains to put them more than four to the mile.'

'Four to the mile, then, all the way to the sea.'

'But that's fifty miles.'

'Yes, two hundred ships.'

'That would be a million tons, but the ships would be light, I suppose?'

'No, no, right down to the load lines.'

'You couldn't see the load lines from the motorway and you don't know that the ships stretched all the way to the sea, but let's pretend that they did—what a convoy! I've never seen such a convoy in my life.'

'Another thing,' said he, 'What about all the other ports? Suppose there were a convoy of that size leaving every big port in Great Britain.'

'That would be about thirty convoys, and still not enough.'

'What do you mean, Grandfather—not enough?'

'Not enough to put in my book. To have enough, let's have a convoy leaving every port in Europe and every Norwegian fiord that can handle five thousand tons of cargo, and Reykjavik and every port on the other side of the Atlantic, so that the man in the moon can see several hundred great snakes with their stupendous cargo, a million tons in each convoy.'

'What is this terribly important cargo?'

'Farmable soil. All the convoys are sailing out to dump it in the sea.'

'Oh Grandfather! I know you're worried about farming and soil, but you're getting rather old. It won't come to that. Why it's a nightmare!'

'Old I am, but not so daft as some I could mention. That picture of convoys of soil is no nightmare; it happens every year already. According to respected estimates, that is the burden carried to the sea by the River Mississippi every year—seven hundred and thirty million tons.'

'Ooh—could they stop it?'

'Quite easily, if enough of the people in the watershed wanted to. As a matter of fact there are government agencies working that way and there are fine schemes in operation and more being advocated, but all they have done is not enough to be noticed. There are great forces working against them, very subtle, tempting forces, the work of the Devil, I should say.'

'There is no Devil,' replied my grandson.

'Well, I'm very glad to hear it.'

However, when I read about the mismanagement of biological resources in John D. Black's book *Biological Resources,* I waver on the subject of the Devil.

I met the United States Conservation Service in 1949 and was shown some fine contour-ploughing that they were doing in the run-down parts of Pennsylvania. There were other activities of various kinds in which a school for boys under rehabilitation from a great city was taking part.

The farmer who was the moving spirit locally gave me a drink of apple juice from his orchard and we talked for an hour or two of many things. He evidently obtained deep satisfaction from what was going on, especially the progress of the boys. It struck me then that his was the attitude that might lighten the burden of the Mississippi river. Ever optimistic, I thought that his spirit would prevail in his great country; still incurably optimistic, I think that it will, but throughout the world I think that the light of learning will have to burn more fiercely than it does. There must be better definition about the problems, a little less of the blurred or comfortable attitudes, and light let into the shady parts.

Let us look at the problem, as good students should, by beginning at the beginning.

Rock to soil: animals

Erosion is a word that we have learned to fear. We have heard of it in many parts of the world and we sense that it strikes at the very basis of life on earth, which indeed it does, as it is not difficult to show. The geologists, however, would not have us fear the word; they teach that erosion is what made the earth as we know it. Without past erosion, the rocks would be bare and we should not exist.

Their meaning is clear to see in mountainous areas such as our nearby Lake District. The photograph of Raven Crag in Kentmere (Fig. 67) shows great lumps of rock that have broken off from recognizable scars during fairly recent times and frost will break them in their turn, as it broke the cliffs above. Water expands by one-eleventh on freezing and cracks the rocks down. The water must get in by an existing crack, which may have come as the rock cooled or dried, or it may have been cracked by earth movements since. The process goes on until the fragments become small enough to form the basis for soil. They form what we call in biology 'subsoil', that is to say, the clay, sand, gravel, stone or even rock, any material that contains no humus, or no more than a little dark stain.

We know from modern experience that such material can be colonized by the one-celled algae that was discussed in Chapter 2, *Pleurococcus*. In the original rocks there were probably more of the somewhat similar algae known by their colour as blue-greens, and which have been observed in modern examples to repeat closely the forms traceable as fossils in Archaean rocks.

We have returned now to the subject of plant succession, which is bound to underlie much of what is written in this chapter. No doubt the algae that we have mentioned would excrete carbon dioxide, making carbonic acid with water, and thus cause some little corrosion of the surface of the rock. The resulting roughness would allow mosses, liverworts, fern prothalli and probably quite early on lichens, so that even the mere surface of the rock would in time carry some vegetation. The corrosion of the surface and this primitive vegetation would help the water to lodge and so assist erosion by frost and make more lodgement for mosses. We know by modern experience that grass seeds can germinate among mosses, which form a little sponge. Mosses form sponge because they store water, but they have no roots of their own and so they are particularly suitable for growing on stone or other relatively impervious material.

---

67 *Erosion in the English Lake District.* Until this first stage of erosion has occurred there can be no soil. Raven Crag, Kentmere, December 1970. (Photo. Geoffrey Berry.)

As time passed, the steep angles of the solid rock gave way to the shallow angles of the fragments in the talus, which is the heap which forms at the foot of a cliff, and there the process is repeated while material goes on descending. The movement stops when the surface arrives at what is called the 'angle of repose' and then the plants can begin to lodge and form a further succession.

This angle of repose is at first temporary, however, depending upon rainfall, wind and sun and when these conditions change throughout the year, as they do in temperate climates, we can see the gaining and losing of vegetation on some of our industrial spoil heaps, as also on mountain screes.

Once we have reached the stage of grass with creeping roots, or stolons, in such kinds as members of the *Agrostis* group, the bents, then the repose is usually nearly permanent as long as the grass is unharmed.

To visualize any of this as a simple succession, however, would be sadly in error. Let us consider for a moment the question of acidity. In the early ages there would not be very much lime compared with what we are used to in lands near the sea. Our soil and vegetation benefit from limestone and from chalk, which are the products of concentration, by marine animals, of the original lime content of the sea, such as it was when it dissolved out of the ancient rocks. To a lesser extent freshwater creatures helped. There was re-deposition of that concentrated lime when the organism died and the limestone shell fell to the bed of the sea to form ooze, which later consolidated as rocks which were raised by movement of the earth's crust, so that we have the limestone hills and the chalk downs.

The formation of limestone will serve as an illustration of what must have been taking place right from the beginning, namely that the first living soil was the product of plants, and of more active organisms feeding on the plants and breaking their substance down to humus-bearing soil. We do not know much about the early stages: there may well have been long ages of slimes before the first invertebrate animals, then vertebrates came to play a part. Even a small caterpillar eating a leaf makes its contribution to soil by its excrement, and before that the algae doubtless provided food for animals as *Pleurococcus* on the bark of trees does for slugs, as mentioned in Chapter 7.

The principles of building fertility, as man does when using agricultural leys described in Chapter 3, must have been in operation from the very beginning long before man came. Vegetation as we see it grows on fertile soil, but the fertile soil grew on the activities of animals over the geological ages, feeding on the original vegetable colonization of the uni-celled organisms, which in their turn were able to get a start because of erosion of the rock at the very beginning.

Soil to rock: animals
When, however, the history is continued, it slowly goes into reverse. When the grass gives way to trees in the way that we have seen in earlier chapters, the soil is replenished from above by the dropping of leaves which rot into humus. Trees also replenish the soil from below because, in breathing, the roots liberate carbon dioxide which with water forms acid to help in breaking down the rocky particles. The simple history then is one of erosion of rock making substrate for plants, which animals eat and so make into soil, but which further erosion by water or wind can lose. Fertility comes up from nothing and can return to nothing very quickly.

The building up process is often stopped and erosion of the soil begun by men felling trees, for building houses, for providing fuel, and in order to clear the land to grow the kind of food that men like.

I think that it may be useful here to stand aside from the main theory and run through a few examples of erosion from far and near, some well known and obvious and some not so obvious.

The Bible kings, Solomon and Hiram, felled timber on Mount Lebanon to build ships for trade and for wood to build the Great Temple in Jerusalem, using a vast number of artificers. The temple soon disappeared, but the devastation of Mount Lebanon was permanent. This is a classic case, and the small piece left of the once great forest of cedars, stands like a jewel on the bare eroded slope of the mountain (see Fig. 68). In this matter we must not, however, overlook the agent that preserved the damage, namely the goat. Dr. E. B. Worthington has kindly written to me about a direct comparison of two areas in cedar forest in Cyprus, a related species, which he visited during the Second World War, and found most striking. In one part, where a zealous forest officer kept the 'black devils' down to a low level, there were hundreds of young cedars on the forest floor; whereas where the goats roamed it was naked of vegetation. Mr. Donald F. Davidson, writing from the British embassy in Beirut, has kindly generalized this for me, from his observations in Cyprus, Turkey and Lebanon. He is sure that, if protected from goat grazing, a felled area would soon be protected by fresh vegetation, including young cedar trees.

The historical case against the goat is truly frightening. Geoffrey

G. Watson in *Fun with Ecology* (not quite the appropriate title in the present context), writes that when St. Helena was discovered in 1502, it was covered with forest. In 1513 the Portuguese introduced goats—and I do not doubt that they also cut down some of the forest—and within two centuries the goats had all but destroyed the vegetation, leaving the soil at the mercy of the weather, so that the island became just a bare crag. Watson also mentions Madagascar, where goats have transformed part of the island into desert. He writes that in 1937 there were a thousand goats and that today they number well over two hundred thousand, so the problem extends into modern times.

Goats attack woodland so effectively that unless on a collar and chain they are a menace wherever they are, a menace to the peasant smallholder and to the agricultural production of the district. When goats are properly controlled, they form a blessing

68   Classical depletion by erosion. Cedars felled by Solomon and Hiram were not replaced naturally because of goats, except for the single grove which has been protected by a monastery wall up to modern times, photo. U.S.D.A.—Soil Conservation Service.

Erosion/I/general notes 185

by milk and by meat, so it is not quite true to say that erosion is due to goats, but rather it is due to unmanaged goats.

Sheep are not often called 'devils', but another classical example of erosion refers to the work of the landowners' syndicate in Spain called the *Mesta*. Their vast flocks of sheep moved seasonally over the landscape clearing off the herbage. The system spread round the Mediterranean lands and is believed to be responsible for leaving it to only dry aromatic vegetation.

Rivington

There is an example of erosion that I can very nearly see from my study window and this introduces one or two other factors. Just above where I live is Rivington Pike which is an outlying spur of the Pennines of an excellent and pleasing shape, fittingly capped by a tower. The tower was built over two hundred years ago for

69 *Modern erosion: neo-pagans on Rivington Pike.* Traditional visitors of the day are further eroding the sacred spot, but the bedrock was detached by young energy on fine days, over the previous year or so.

an owner who wanted it for a summerhouse and shooting box for himself and his guests. The elevation is something over 1100 feet. The Pike and the tower have formed an important landmark for a very long time and when I look towards the Pike from my fields on Good Friday, the skyline is black with people whom I tend to think of as neo-pagans (Fig. 69). They happily confess that they must go up Rivington Pike on Good Friday, but none of them can say why. The tower has even been reproduced in pottery.

Of late years, boys have climbed up inside the tower and thrown some of it down, thus spoiling the parapet. The landlord, rather naturally, thought that it would be easier to have it pulled down than to keep it built up against the vandalism. When he proposed to do so, however, there was such a protest that he desisted. Sons and daughters of South Lancashire wrote from far and near that it should be preserved. It was made quite clear that the Pike and the tower, which are visible for about thirty miles, over an arc of half the landscape, is something of real significance, if difficult to define.

About seven years ago, I rode up to have a look at the state of it, and found that the threshold stone of the door was several feet up from the rock. The foundation stones were on exposed bedrock, which slopes naturally, and were surrounded by a bed of loose stones. This stony bed was continued as a gully down the side of the hill, some three or four feet wide, also filled with loose small stones. The tower had a flat roof from which the drainage appears to have fallen through ports left in the parapet at the corners. At the foot of the foundation stones there are remains of peat, showing that for a long time heather grew at its foot, although it was some feet below ground level. The falling roof-water may have eroded a ditch or, as seems likely, owners dug a ditch to lead the water away and down the slopes. Some erosion would result, but not seriously so for about two hundred years. Then came a vast increase of visiting public, which at first did not damage the beautiful fine turf, well cropped by sheep.

In October 1969, before finishing this chapter, I thought that I should examine the tower again and rode up to do so. Before I reached the final hill I saw that the gully had now become very much wider, some fifteen feet or more, and it was no longer filled with round glacial stones but had eroded so as to expose the bedrock, resting on sand. Seven years before neither rock nor sand had been visible. The rock consisted of large cuboid blocks, lying in their natural bedding no doubt, with about six courses of the strata visible, looking like giant's walling, and two blocks detached and some way down the sandy slope. When I reached the tower I found that there were none of the small stones that had made a bed round it, but instead a flat yard formed of the bedrock.

Not noticing the significance of the cubes of bedrock half way down the slope of the scar, I thought that it would be a very long time before the bedrock became sufficiently eroded for the tower to fall, perhaps in three or four centuries I thought. But I was wrong; it may fall much sooner because energetic boys have found that they can use iron fencing pales as crow-bars to dislodge the bedrock; and within weeks the giant walling had gone and no longer showed in pictures in my course (Fig. 69).

Here, then, is some erosion due not to animals mismanaged but to human non-management. Nearly all the small stones are gone from the summit, and there is no mystery in that. In a few minutes spent in watching young visitors you may see odd stones sought from under the edge of the remaining turf and hurled into the void below. Those in the gully would also be dislodged by intrepid mountaineers, who despise the adults' easy detour up the grass slopes.

So, Rivington Pike is eroded by people, possibly helped by sheep, but mainly by people, and the scar, deplorable to all who see it from near or far, bids fair to grow until the beauty of even the distant Pike is destroyed. It is odd that water from a small roof should set going the great erosion of a landscape feature.

In front of the Grammar School at Rivington evidence of erosion by a sunken road is presented by an oak tree, elevated on a cone of supporting roots. I showed a picture of it in Chapter 7 (Fig. 34).

Before leaving Rivington we may look at other examples. The fields that I rent there are nearly flat, and therefore lie wet when there is a great deal of rain. During two or three successive years which were very wet, there was a considerable increase in the overtrodden areas round gateways, which are, as it is called, 'poached', that is, there are footholes full of water which turn into mud. Had that land been sloping but with flattish places on it for the cattle to poach, the resulting mud would have washed away. In that way, cattle may poach areas, especially in countries where there are too many kept because they are a form of wealth or prestige. Even when few cows are kept their damage is seen round gates or watering places; especially if out in winter, when the grass can hardly grow.

Kersal Moor
For erosion due to plain overtreading by people we may instance Kersal Moor in Salford, where the class carried out its practical work. Its sandy hills are barely covered with thin hair grass and bents characteristic of sandy and acid soil. Mr. Pearson, the custodian, tells that the higher of the hills is losing height over the years. The sand is moving down to a lower level where a lower hill is growing. His evidence seems to indicate that particle erosion is taking place through that thin grass without the grass being totally abolished. It cannot, however, grow very rich or thick. This happens to suit the function of the place as a rough recreation spot, and to cure the erosion might not be beneficial. Not all of the vegetation of the moor is of that thin type; there are some patches of big, wild coarse grass, namely cocksfoot, and with it there is associated wild white clover. This sends out runners like a strawberry plant and so works against erosion. Soil samples taken by the class show that there is a higher pH where there is clover and cocksfoot. Sometimes the cause is obvious, in that the patches are near walls or buildings, and we can reasonably assume that the higher pH was due to limey material having leached from the building works at some time and having been maintained in the soil since. Where the patches are far away from walls we must seek another cause and this may be the large bonfires that are built and burnt in November. They would appear to be very damaging to the grass and doubtless they are so for the time being, but I have very little doubt that the high pH near the clover plots there could come from the ashes of all the timber that has been burnt.

The moor would not look tidy if there were cocksfoot all over it. It would need to be cut, and then it would look like a well-trimmed park, which is not what the inhabitants really need there. If it were not regularly cut it could catch more litter than it does at present and not look well. At present it has a rather wild look about it, which is probably of inestimable value, to the children of the neighbourhood at least, in the City of Salford.

Erosion of Kersal Moor keeps the fertility cycle going slowly and renders the land of less agricultural value than it might be, which happens to suit the requirements of the place, but this may not last. I have mentioned that Mr. Pearson told me that its hillocks are becoming lower. If the moor went too far towards non-fertility, it would blow about in high hot winds, and that reminds us of the disastrous dust bowl of North America, a situation immortalized in John Steinbeck's novel, *The Grapes of Wrath*.

Wind erosion

A classical story of that area was that a banker from the East went to the Middle West to find out why he was not receiving any payments on his farm mortgage. The farmer met him at the railway station and the banker asked to be taken to the farm, to which the farmer replied that it would not be necessary, if he would only look up in the sky he would see the farm coming towards him.

That is wind erosion, one of the more conspicuous cases. There must have been a great deal of it in the past to build up the deep loess soil that we find in the centres of continents. In America they farmed that soil, blown there in past ages; it was so deep and rich that everyone said it was like a goldmine, and I remember it being described to me in that way when I was a boy. It ran out as goldmines do.

Detachment of mat

The problem of erosion is more severe on steep slopes. There is a story from Africa, told to me by the late Victor Van Straelen of Brussels, who was one of the most active naturalists interested in conservation anywhere. People in some parts of Africa were farming on the side of a mountain and every year were cultivating the soil, but it was slipping down and so they moved down after it. The Western experts who came said that in twenty years time no more farming would be possible on the mountain. The local government was persuaded to insist on the African farmers terracing their land and growing elephant grass to hold it. Elephant grass is a great grass growing six feet high or more. The mat formed by the elephant grass roots, however, proved to be such an attractive home for all the little voles, to say nothing of beetles and all the animals that feed on beetles, that those miners completely detached the turf from the substrate. Then came storms and the whole lot of the turf went down in large sheets. Instead of the soil leaving the mountain in twenty years, it left in one year.

Unfortunately, my enquiries since Van Straelen's death have not revealed where this took place; his colleagues in Brussels are prepared to exclude the Congo. The example will serve, however, even if it be merely scientists' gossip, which I do not think for a moment because Van Straelen was an exact man, to show that erosion in each case is a complex of technical, social and political issues and nobody from another country should rush into it. It is not a matter to be dealt with by pure theory, nor even by careful synecology alone without macro-ecology, and if there is no history to go on there should always be some form of trial on a pilot scale.

More about trees

The descent of the elephant grass is a reminder that in order to stabilize land there should be roots of trees or bushes as well as those of grass. The value of trees and bushes is widely understood among conservation-minded people; indeed, it is sometimes spoken of too highly. In order to emphasize the value of trees, I am afraid that I myself have been guilty of suggesting that the land will be safe if it is under trees, whereas, strictly speaking, one ought to say that it would be safer for a decade or two, so long as the trees do not grow too tall.

We must remember that foresters have told us that trees vary in their qualities of soil improvement, although all of them improve the soil at first, until they form a canopy.

If we turn to the wood which I tend to use for my students, which is a piece of mainly natural regeneration on an old spoil tip, it is worth mentioning that in 1960–1 I could not find the colliery shale without digging for it. I had to go down six inches or so under soil and grass among the trees. But now in 1969 the shale shows up so that when I am showing it to people the tip is quite marked, as it originally was. The shale went through its ninety-year natural regeneration until it was growing quite reasonable timber, then the canopy came and killed the grass, with the help of treading by the public, and then erosion set in again.

Thus, if grass needs to be reinforced by trees in order to be safe cover for soil, it may be equally true that trees need to be supported by grass. Indeed, if we go into any mature coniferous wood that happens to be on a slope we are likely to see signs of the soil eroding between the trees, and roots exposed.

So, trees do not form a safeguard automatically, and there is still much to be learned about their conservation value. This was shown to me clearly by Dr. Frankland of the Nature Conservancy's station at Merlewood, near Grange-over-Sands in North Lancashire. In Roudsea wood we saw brashings and waste timber left to form humus on the forest floor, all in trial plots with controls. There were differences in floor cover caused by roe deer, that is, whether they were fenced against or not. There was some rock under old yews which was strikingly bare, except where a seedling grew in a crevice which the deer could not reach.

Wherever erosion appears, it is as well to ask what animals frequent the place. At Kersal Moor there are rabbits, probably an appreciable number, because there are also foxes, of which Dr. Parr tells me he has heard the unmistakable vixen's bark—not the animal most people would expect within Salford's city boundary.

Savory on Rhodesia

The havoc that the grazers and browsers may cause has been illustrated by C. A. R. Savory (1969). He writes about the national parks of Rhodesia and gives what seems to be very sensational and very bad news.

I used to make a rule never to refer in a book to anything so temporary as a scientific paper, unless it had been published for at least five years. However, Savory's paper must be included here I think, with the proviso that it may not be confirmed in its entirety; but equally it may.

It will be remembered that in Chapter 2 we considered the styles and power of grazing of different domestic animals. The cow, and the other bovidae to some extent, need grass which is long enough for them to get hold of between the tongue and the hard pad of the roof of the mouth, because they have no upper incisor teeth. Some of the bovidae, however, such as the sheep or goat, have a split lip and can get the pad closer to the ground, so that sheep can graze much closer than cows. The rabbit has upper incisor teeth and a split lip so that it can graze closest of all.

It follows that if by any chance the long grass has gone completely from a grazing, and only shorter species survive, the sheep will out-compete the cow.

That is the sort of thing that Savory believes has happened in the national parks of Rhodesia where scores of rhinoceros and other long grass feeders have been found dead from starvation. By prohibiting all killing of rhinoceros, for benevolent reasons, the numbers were allowed to increase to such an extent that the perennial grasses on which they relied died out and their place was taken by annual grasses, which unlike perennials have to rely entirely on the current rainfall. The perennial grasses with their rootstocks had not been so dependent. Consequently there were times in some seasons, while the annual grasses were not yet grown, when there was no long grass, and there were even times when there were bare patches in parts of the range. From the bare patches there would be more evaporation and so the water-tables fell and the lush grass had no place to grow. At this stage the range is reasonably suitable for a closer grazing animal like the impala antelope, but if the antelope is allowed to increase to unlimited numbers it will eat down the shrubs on which it feeds until they too are over-grazed and eaten out; then the trees die. The final stage is that the range goes to desert. Thus even a range that starts with thick forest is not safe if the animals are mismanaged. A newspaper supplement on Zambia (*The Times*,

October 1969) described new policy there, including the controlled slaughter of big herbivores.

It is a very strange thing that national parks, which are intended to be game-reserves, can by over-preserving the game be destroyed by erosion, or so it seems if Savory's account is anything like sound. I am sure that it is at least partly right, being a development of the appreciation started by Leopold and reported in Chapter 9. One aspect of that, which was supported by other evidence in that chapter, is the necessity of the predator, with its therapeutic value in maintaining the population of the prey at a reasonable size and thinning out the weaklings.

Mismanagement of animals

If any student wishes to know more about the subject of erosion, he should study the publications of the U.S. Conservation Service and one or two excellent books that have come from America, but even the notes that we have put together here do enable us to say with some confidence that erosion chiefly arises from mismanagement of animals, which has cropped up in almost all of the examples given, especially if we include mankind mismanaging itself.

The most common type of mismanagement is what has been called farming to leave, a phrase dating from disaffection between landlords and tenants; this takes us back again to Chapter 4. All who have studied the subject do, I think, agree that it is the cause of loss of humus from the soil. Soil with sufficient humus does not easily blow away or wash away, although there must be some areas where it would do so unless protected by shrubs.

This is John D. Black's conclusion from America, where the problem is of the enormous scale illustrated by the convoys of ships with which this chapter opened—but the Mississippi watershed is not the worst over there. The title of his book begins, *The Management and Conservation* . . ., but it might equally have been in the negative. Even the dust storms in England, of which we spoke in Chapter 4, would be ascribed without any doubt by the practical men in the Farmers' Club to attempting to farm without animals and therefore not requiring hedges. Many official advisers who in the past have helped to make farming-to-leave fashionable are now seeing their former error.

It is beginning to be sensed that farming without humus has been due to an attitude of 'don't know', which, to be truthful, should be joined by one of 'don't want to know'. Human greed precluded human knowledge until the land was so run down that bad farming became a necessity in order to remain in business at all.

I claimed at the beginning of this chapter that the burder of soil in the Mississippi water could be reduced quite easily if sufficient people living in the watershed wanted to do so. The fault is in the hearts of men and erosion will continue until it is realized that no man should have the impertinence to make a profit by leaving the land in a worse state than he found it. Putting speculations aside, there is no known way of avoiding land robbery, except by the keeping and right managing of animals, which leads me to the particular study of the next chapter which is concerned with one form of erosion, that of river banks. Probably it is worth a whole chapter, not so much for its intrinsic importance as for a further example of the methods of investigating ecological problems.

13 Erosion/II/swift rivers

History

In the last chapter, I said a little about the erosion of Rivington Tower, which is within reach of a great many people who go that way for recreation. The same is true of Borrowdale in the Lake District.

The hypothesis for Borrowdale was founded on what I have called in Chapter 10 an 'awkward fact', that is, I noticed something that was not in keeping with the popular views of how trees grow. Most people think that trees are planted, having first been reared in a nursery, and that method certainly is the way to grow good timber, giving it a good start for a straight trunk free from large knots. I think that most people would tend to say 'planting' trees, rather than 'sowing' trees, or even trees growing of themselves. But, we have already mentioned at the beginning of this book the standard case at Rothamsted where a field turned into a wood without anybody sowing or planting anything.

It was a similar phenomenon that I saw nearly thirty years ago, when I was camping near a miller's weir at Newby Bridge, a mile below the foot of Windermere. I often looked across the mill leet at some nice trees growing on the land between the leet and the river. The land was principally of stones. It occured to me then, that stock would not reach such a place because it was surrounded by water with no good bridge for them, and it was obvious to me that the stones were held together by the roots of trees in a thick vegetation of ash, alder and sycamore. That the trees had clearly not been planted there is an observation that I think can possibly qualify as one of these valuable awkward facts I have already mentioned.

From that time onwards, I noticed in the Lake District that where there was complete protection of a river bank from stock there was vigorous growth of at least those kinds of trees I have named, and sometimes others. The islands formed between mill leets and rivers, although sometimes low-lying and frequently flooded, always seemed perfectly safe and never disturbed by the turbulent water that went rushing by, nor by anything that it carried with it. I therefore formed the hypothesis that the stability of Lake District river banks was often due to the growth of trees whose roots held the stones together. Where there were no trees, I thought that stock had eaten down promising seedlings.

It happened that at that time I went to see my brother, the late R. B. Graham, who owned Chapel Farm in Borrowdale. As headmaster of Bradford Grammar School, he had been looking out for come constructive work of national importance for his boys to do and I found him building a wall on the banks of the River Derwent across a small bay which the river had made in his fields, to his very considerable distress. He was building the wall very solidly with sufficient tehnical help, as well as his boys' assistance from time to time, and he was receiving a grant and respectable advice from the Ministry of Agriculture and from other authorities who were in favour of what he was doing.

Having for many years watched sea walls broken into separate lumps, at Corton for example near Lowestoft, I was not very enthusiastic about the wall, strong as it was in itself, but it was too late for a younger brother to intervene with advice. I did quietly ask a local farmer whether he thought the wall would stand and he replied that he was dubious. Remembering Newby Bridge, I thought it would have been much more effective if he had merely fenced the edge of the devastated area and the rest of the river banks in order to prevent the stock reaching them. I opined that wrack would collect among the round stones that formed the floor of the little bay and that trees would grow in it, provided that they were protected.

I visited Borrowdale again in 1967. In the previous late summer there had been two rainstorms of a magnitude not experienced before in this century. They did more damage than any that people could remember with certainty, although there was some dim recollection that there had been two storms in the later part of the last century that were nearly as bad.

Stones, some the size of boulders, rolled down the watercourses, blocking the river and banks and so raising great floods. The flood in the Scafell Hotel at Rosthwaite rose about three feet in two hours. One of the guests, Sybil Forsyth, who was in a chalet in the garden of the hotel, was swept down the main road of the village and battered and bruised. She was rescued by a local man who was standing in the stability of a hedge, who managed to grab her as she was swept by. Her account of it in *The Countryman* (1967) is well worth re-reading.

In one of the fields at Chapel Farm were spread what now looks like two or three hundred tons of these round stones. The farmer had pushed them away into a bank, and my brother's widow asked me to come and look and advise her because she wanted to plant the bank with trees in order to improve the appearance of the valley. I told her in correspondence that she would have no difficulty if she could arrange with Joseph Weir, the present owner of the farm, to fence it sufficiently against sheep and cattle, then it would vegetate itself, and for some time I resisted the call to go to Borrowdale and have a look at it. I am very glad that I finally gave in because the whole matter turned out to be very interesting.

On arrival, I went first to look at Dick's wall and found that it had been breached—my sister-in-law told me within ten years of its being built. It had apparently been broken at one end and water entering had done more damage to the land at both ends. This lies just downstream of Folly Bridge, which Dick had also repaired in his day. I started then to look all round to see what had happened—thus making the part of the enquiry that I would call the survey.

Observations in Borrowdale

Let us first take Dick's wall and notice that the middle piece, which is still standing, has a protective screen between it and the river. This screen is formed of young alder trees which have seeded themselves and obtained a lodgement since the wall was built, as Joseph Weir confirms. At the downstream end of the wall, the water has scoured its way in and exposed a fresh bank to the field behind, showing the turf and soil in the field to rest on a bed of packed and rounded stones. The bank had been intact when my brother ended his wall there and the river was quite straight, so there was nothing obvious for the water to impinge on. This and all other observations suggest that the problem is that the water eats its way into the bank rather than batters it with boulders, which is how the local people tended to describe the effect.

At the upstream end of the wall there is a cement plinth made at the foot of the wall, now more than a foot up in the air, showing undermining rather than battering.

Let us take now the observations that can be made going downstream from Dick's wall.

First we find two stumps of trees in the river, a yard or two away from the bank (Fig. 70). These are quite dead, but if we look across to the other side of the river we see similar stumps which are still alive although cut down. They look as if they had once been laid horizontally. Then turning our gaze upstream and still

Erosion/II/swift rivers 197

70 The Borrowdale Derwent *(left)*. Dead stumps are standing off from the near bank which is eroded; there are live stumps on the far bank, which is intact and was formerly protected by a wall, now a fallen remnant. On the far left young alders grow on the near bank only where it is protected by Dick's wall.   **71 Tell-tale stump** *(right)*. This is alive on the beck side but dead on the field side, and new twigs on top, which had been out of reach of sheep, were cropped by cattle when let into the field.

looking across to the left bank, we find that there are quite a number of these live stumps and lying behind them the remains of an old wall parallel with the river. Surely, this wall once separated the stock from the river bank. This does suggest that there was a live barrier, in the form of coppice trees, to any aggression of the water on that bank, but now the sheep are all about near the stumps, which probably was not so when the ancient wall was standing. Proceeding then a little further downstream we come to a tributary called Coombe Beck and if we turn up it to where the road bridge crosses, we see a most remarkable stump (Fig. 71). It also lies horizontally, and on the beck side it has many live shoots, but on top the shoots have been eaten away by stock, while on the field side, which is entirely open to stock, the stump is dead and carrying fungus. This sight can be seen quite easily from the road and is most impressive. On our first visit, there had been only sheep in the field, and they had not then reached the twigs on top, which were flourishing sycamore shoots, but on my second visit I saw that cows had been allowed into the field and they had reached over and eaten the top shoots down so that they were only short twigs with no leaves on them.

Before leaving that field to ascend Coombe Beck let us retrace our steps and go up the River Derwent a little towards Folly Bridge. About a hundred yards upstream from the old wall is a piece of bank on the opposite side that Dick reinforced with stones and iron stanchions, the mass built as a slope, throwing a

72 *Coombe Beck protected* (*left*). Looking *upstream* from the farm bridge, the bank is stable. Moss-covered stones and tree roots hold it, and the tree is protected by a wall behind it and a fence on the bridge. 73 *Coombe Beck devastated* (*right*). Looking *downstream* from the same bridge, bare loose stones are in the beck. Tree protection has been lost through past felling, shown by the 'foot-rest'. It was not restored because that bank was open to cattle and sheep.

few bags of cement on top. He thus left a solid sloping reinforcement that stood unharmed for twenty-five years, but it was open to stock and in 1967 it had just begun to loosen. On the near or right bank there was also to be noted an isolated fifteen-foot ash which had swayed in the wind and loosened the bank.

Leaving the main river and going up Coombe Beck, the field road leads to a farm bridge belonging to Chapel Farm, on the way to an old mill. Standing on the bridge one looks up at an extremely well protected part where the bank stones are all in place and green and mossy (Fig. 72); evidently they have not moved in spite of the flood with its devastating effects elsewhere. They are in fact held by roots of trees which had been protected, until they had grown to a considerable size, by a field wall built behind them and by a fence on the bridge in front. That was, however, quite untypical of Coombe Beck. Still standing on the bridge but looking downstream, one sees a complete contrast (Fig. 73). Upstream there is rural beauty, downstream there is a devastation of white stones, all of which had evidently been moved about in bank repairs. One wonders why there should be this difference. It is not difficult to see an answer, because there are four dead stools of substantial trees that were evidently cut down for timber many years ago and from which the stock has not been fenced off. So that bank had no live roots.

Going upstream from the bridge fifty yards or so, one comes to the old mill. The floods had dislodged the iron mill wheel, knocking or sucking down the wall that supported its bearing on the stream side. The wooden bearing with its hub was lying on the opposite side of the stream but the iron wheel had been propped up again with stones by someone who was clearing the stream afterwards. The supporting wall for the wheel was also the retaining wall of the leet and it had been partly built up. It was easy to distinguish the repaired part from the part that had not been damaged because the stones that had not moved still had moss on them, which they had grown in the course of time. It was evident that the retaining wall had been protected by a large alder tree which was quite undisturbed, but there was still another significant specimen, or so it seemed, namely the stump of a sycamore on the breast of the wall, the part that was jutting most out into the river. This stump had been browsed by sheep reaching down from above but it was still alive, as is shown in Fig. 74.

Proceeding upstream again, one comes to the tanks of the West Cumberland Water Board, and just above them is a very large cavitation on the right bank which is made of the same soft glacial soil as the valley bottom but with the boulders not packed quite so tightly (Fig. 75). They are bigger, however, and there were a good many like them in the bed of the stream. Going further upstream, more cavitations can be found but the big cavitation behind the reservoir was particularly interesting because it had evidently expanded until it was stopped by some small alder

74 Coombe Beck mill. The flood dislodged the mill wheel, which has been propped up again. In the wall of the leet there still stand some moss-bearing stones, apparently held together by the roots of the sycamore stump, which is alive although attacked from above by sheep.

75 *Coombe Beck: cavitation in sheep run.* The banks are of soft glacial soil and sheep prevent the natural replacement of trees. Small trees such as sallow and thorn or rowan would help; but should always be polled at a height of twelve feet, or laid as at Kentmere, shown in Fig. 77.

and rowan trees, which were holding the stones together. They were doubtless helped by the roots of a fine ash that is to be seen further up the slope.

Observations in Kentmere: major discussion
While making these observations in Borrowdale, it occurred to me that it would be valuable if we could find some other valley in the Lake District served by a swift river and investigate the treatment given to the river by the people there. I thought of Kentmere, which is famous for the speed and floods of the River Kent, and so my companion, keen Lancashire naturalist Tom Creear, and I went there.

One thing that Tom found gave us both considerable pleasure, and that was a comparison of two ferns growing on the river bank (Fig. 76). One had been completely flattened by a flood four or five days previously but the other, although it was growing nearer the water, was standing upright and untouched.

The only explanation that we could find was that a small branch of a wild rose was lying in the water above the fern that was standing up. It seemed that that little brake on the speed of the water had made all the difference. Tom put a stick in the water and paced it from the bank and found that the speed was six miles per hour. Doubtless it was greater during the flood.

The ferns were on the stretch of river below Hartrigg Farm, which is the highest farm in Kentmere, and the river there seemed to be quite under control between built-up stone banks fringed

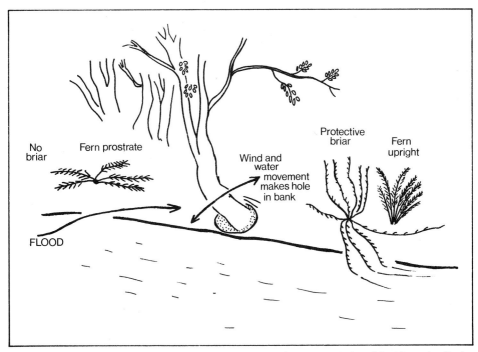

76 Some complications on banks. Dr. Stanley Frost drew this as a reminder of significant detail. The two ferns in Kentmere are drawn, and the protective effect of a mere briar acting as a flood gate; together with damage in the holding effect of trees if they are isolated and can sway in wind and water.

with short trees which are protected from stock by fence or wall.

When we started to walk downstream we found a place where the fringing trees stopped and the stock could reach the bank, and sure enough, within twenty yards there was a small cavitation and one or two of the bank walling stones were down in the river. At the downstream end of the same field there was another such cavitation and then the trees began again and there was no serious cavitation until we got to the neighbourhood of the foot-bridge. Looking up from the foot-bridge we saw a tree which had fallen across the river and the disturbance had lifted a large bank walling stone so that it was resting on the root of the tree. The water would undoubtedly go round the tree and bring down more of the bank walling and probably nearby fencing.

Forty yards or so below the foot-bridge there is an isolated ash tree, no more than twenty feet high, which is now swaying to such an extent that one could stand in the cavity it had made in the bank.

Further downstream we come to the road bridge from which the small sized trees can be seen fringing both banks of the river, and, entering a nearby field we find that the river has been made to turn a whole right angle without tearing away the field (Fig. 77). It is constrained by built-up walls of big stones which have been protected by allowing ashes and other trees to grow up to about

77 *The method at Kentmere.* The swift flooding Kent is tamed in Mr. Salkeld's meadow (below the village) where it is made to turn a right angle. It is confined by trees on each side. These are laid in hedgerow fashion and are too massive for stock to kill them by browsing. It is, however, only fair to add that this is a meadow, which is less severe on live fencing than most pasture, being shut away from stock in the main grazing season.

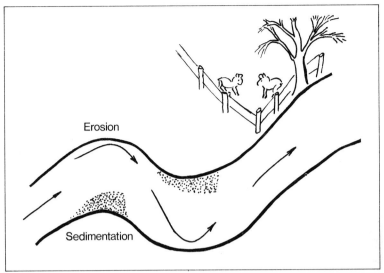

78 Forces on river banks. Dr. Stanley Frost drew this summary of the work in Borrowdale. The general displacement of a river bed in soft soil is shown on the left. However, as illustrated on the right, it cannot take place where tree roots hold the bank, and the tree roots are present because sheep have not been allowed to destroy the seedlings.

twelve feet high and then laying them to form a very thick hedge. It is only fair to say that this is meadowland, not pasture, but stock is there to graze the aftermath and for all I know may be there in winter as well when animals are particularly keen to eat treebark. This hedge is so substantial, with horizontal pieces as thick as a man's arm or thigh, that no stock would kill it out of existence by winter browsing.

The evidence I have detailed all points in favour of the hypothesis I hold: bank stones stood fast where they were held by the roots of trees and trees had grown where the banks were protected from stock.

My theory can be fortified from macro-ecology by noticing what is growing now on the fells and considering the question of trees and their regeneration.

On the slopes there are a number of isolated trees, rowans, thorns, occasional common ashes and alders, or in some districts, scrubby oaks. Trees as big as that would not take any harm from stock, but sheep and cattle prevent replacements either by coppice shoots or by seedlings, for they are very fond of both these forms of tree growth. Thus the banks become bare in the course of time but gradually, so that nobody would notice that the stones rolling down the tributaries were becoming more and more frequent and troublesome.

Tentative advice based on the theory

When pieces of landscape are giving trouble of one sort or another, it is often justifiable to try a theory, on a small scale at first. I consider the theory here sufficiently well founded for a certain amount of advice to be proferred, but always in favour of a pilot scheme if the delay can be tolerated.

Were I facing the problem which my late brother tackled, and which many landlords and authorities do face today, I should be inclined to fence a strip seven yards wide along all soft banks and to repair current cavitations on the lines of Dick's sloping bank, not of his wall. I should sow hay seeds on the slope to take some hold and seeds of ash, alder or sycamore if available, or acorns in oak country. At especially vulnerable points I should drive in pegs of sallow about a foot apart along the bank. It might be sufficient, as Ken Dawson of Hartrigg Farm suggested, merely to lay sallow branches, which root at many joints in some wrack close to the water's edge, but they would probably need reinforcement by the deeper rooting kinds of trees that I have mentioned. I should never, I think, use the iron stanchions that my brother used, but would substitute stakes of willow, even of sallow, most of which I am sure would live and hold the stones together until the deeper rooted trees got hold. To prevent loosening by swaying or falling, I should top or lay everything resembling a tree on its reaching twelve feet to fifteen feet high and would hope to get rid of the anti-stock wire fencing in the course of perhaps ten or fifteen years. I came to the conclusion also that river engineers did need warning that if they straightened channels they could expect the water to go faster and so any flood damage would be increased. Just as the rose branch saved the fern, and just as the flood gates saved the abutments of bridges, there is something to be said for delaying the water on its downward way in certain circumstances, which water supply engineers are reluctant to do because the slower the water goes the more they lose by evaporation.

It may be mentioned here, that large erosion of river banks has been checked in emergency by facing with old motor cars in Cumberland, and by laying great mats of linked old tyres in Minnesota. In both cases, it is intended to grow a tree in each element of the structure.

Turning back to the framework we have used in this book, the autecology of the fringing tree species and the synecology of trees and sheep and the macro-ecology of the history have all played their part, although there has been no attempt to keep them separate in this chapter.

At this point we could remember the beaver, who intervened in Chapter 6. His dams slow the current down and his logging operations prevent trees near the waterways becoming large enough to sway and so damage the banks. However, the beaver's work may or may not fit in with other schemes.

After a summary of the work in the Lake District was published in *The Countryman* (1968), I received a letter from Frank Law the engineer to the Fylde Water Board, Blackpool, Lancs., which owns the Stocks reservoir and a very considerable amount of high river catchment in the Forest of Bowland. Later he sent me an account of flood damage done there almost exactly a year after that in Borrowdale. It was very similar to the Borrowdale damage but over a much larger area, as described in the Board's annual report for 1967–8. He invited me to visit his area. The moment we set foot in Bowland, we saw an object lesson of the new approach to river erosion. This was at Bishop's House on the banks of the river Dunsop.

On the new land that had been built up between the house and the river bank to take the place of what had been scoured away, there were some thousands of sycamore seedlings, nicely spaced at about a foot apart, in their fourth leaf stage and looking very flourishing. Unquestionably, there was in embryo a protective piece of woodland between the house and the river, with nothing to prevent it from growing except the sheep. The sheep were to be seen round about with their lambs, eating grass. Everyone would agree that in the winter, if not before, the sheep would eat those sycamores away. But for this hazard from the sheep, nature had, by sowing from the mother trees that formed the wind shelter to the house, provided an ideal protection against any flood disaster such as that suffered nearly two years before.

I am glad to say that Frank Law agreed that the patch should be fenced to serve as an object lesson for the future. He also intends to fence the banks and generally as the Borrowdale work suggested. Thus there seemed to be a prospect of the theory being raised into something like a law, if you can give it that dignified name, through the practical application to be made by Frank Law, who is an ecologically minded engineer. He has already to his credit a fine experiment on the interception of rain by trees, which leads to the conclusion that forestry, at least by the conifers favoured by present fashion, is not a conservation remedy for use extensively by water undertakings. But to fence water courses, leaving controlled drinking places for sheep and stock, and to grow live fences on the Kentmere model, seems likely to be given a good trial.

Our next chapter, on another practical matter, continues the application of science in the open-air to another topic in which erosion is again a key factor.

Reclamation of spoil lands/I/technical 14

Something immediate
When I started in fishery science, an aunt said, 'What a strange way of making a living,' but of course she could not be expected to know that the smacksmen had demanded it. I do not think that either the smacksmen or I myself expected fruition of the researches to be so long deferred. In the 1890s they were disappointed that no scheme of conservation came of their efforts, and in the 1940s I suffered a repetition of their dismay.

I thought that it would be after my time, when international agreements sharing out fishing power became effective practical politics.

That this should not happen sooner was a great disappointment and it seemed to me that I had not worked very effectively during my professional life. I was therefore delighted when, on nearing retiring age, I encountered a problem that could be dealt with in a matter of two or three years, or at any rate set well on its way towards contributing to visible improvement of the landscape.

The discrepant, or awkward, fact that set the business going was that at the foot of slopes of spoil, commonly called slag heaps, there was often grass growing. This should not have been possible according to the common gossip that the material was either barren or poisonous. It could not therefore be the nature of the material that was preventing grass from growing on the slopes, sometimes for ninety years and usually for about thirty years, so it must be something physical. The slope suggested erosion. If so, the measures I had seen in Pennsylvania in 1949, mentioned in Chapter 12, should enable grass to grow all over the heaps.

When I called on him, I found that the County Planning Officer for Lancashire had got so far as growing grass direct on shale through having had the good sense to make experiments in flower pots, and, by the way, I still recommend that anybody engaging in this work on any tip should start by a series of flower pot experiments. They are very easy and give positive results very quickly if one uses such a plant as mustard.

Already in 1955 the Forestry Commission had issued a paper by

R. F. Wood and J. V. Thirgood giving useful advice on tree planting derived from an extensive survey.

Practical trials

The County Planning staff showed me their experimental plot at the Bickerstaffe tip where they had cocksfoot grass growing along ridges that they had unintentionally made when walking to and fro on the face of the tip. J. K. Brierley (1956) had concluded from a preliminary survey of about forty tips that mechanical factors rather than chemical probably prevented vegetation growing on them, and he developed his ideas to the extent of showing the terracing effected by wavy hair grass and colonization by birch seedlings on the hair grass turves. Brierley's diagram showing this is copied here (Fig. 79).

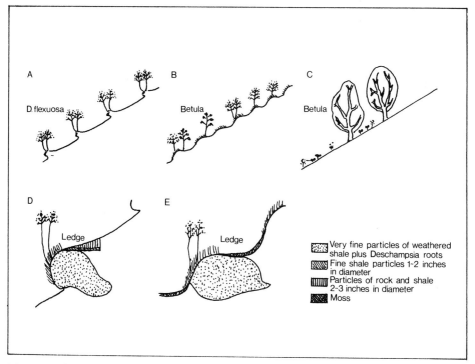

79 *Natural regeneration.* After Brierley (1956) with permission. (A), (B) and (C) are successive stages and (D) and (E) typical early stages in natural regeneration, on a larger scale. Brierley's version of the order of vegetation is thus: wavy hair grass terrace; moss and birch. The present author agrees, but believes that there are also preparatory stages: fleeting invasions of green alga, *Pleurococcus*; then moss, hair grass and the rest of the Brierley series as pictured.

I. G. Hall's paper, which is an extremely useful one from the botanical point of view, came out in 1957 and J. A. Richardson's in the same year. It follows that in 1958 it was really only necessary to confirm these comparatively recently published observations and to generalize them, so that any man living near a tip, if he were annoyed by its appearance or by the dirt that the children carried into the house from it, could go and grass it over and in that way deal with both troubles.

I received a grant from H.M. Development Commission administered by the Community Council of Lancashire. which had itself made some trials to mitigate the idleness of the Great Depression. The Council circulated interim reports of my results. A rather fuller one was issued in December 1961, called 'Nature Repairs' which I have placed in the library of the University of Salford. There are three or four other copies in existence, namely in the libraries of the Ministry of Housing and Local Government and of Agriculture, Fisheries and Food, and that provides perhaps sufficient reference for anyone interested enough to read up the subject in a critical way. Five hundred copies of a summary were also issued. Here I shall try to give a general account of the work, as an example of ecology being put into practice.

The County Planning Officer had reported some trials recently made at Bickershaw on which he had found it necessary to use limestone at a rate of about six tons per acre. I therefore added to what I had observed about erosion the idea that there would have to be considerable dressings of lime. I ascribed these to the probable presence of sulphur, but later Philip St. J. Edwards who took over this work at Salford, showed me that the sulphur had, in some cases at least, disappeared, but had left behind it sufficient acid to require the lime. In one instance we found that it was a single dressing that had been sufficient. The ochre deposits, which I had been seeing in shale and along with other practical men had thought to be sulphur, were, Philip Edwards found, formed of iron oxide left after the sulphurous iron pyrites had been dissolved away by the water and carbon dioxide of the air. I should add that Lancashire shales, demanding six tons, were worse than some others, namely Yorkshire shales, whilst in Nottinghamshire there seems to be no need for any lime at all.

I must make it plain, however, that I did not attempt to solve or follow up the chemistry of the pit heaps, I simple confined myself to the practical application of what had been found out already by Brierley and by the County Planning Authorities, especially those of Lancashire in their trials at Bickerstaffe and Bickershaw.

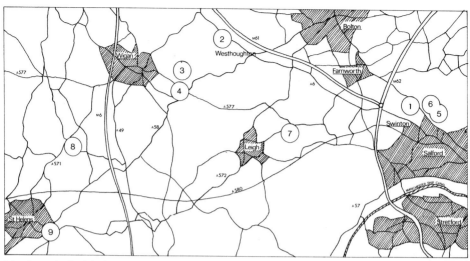

80 Spread of trial plots over South Lancashire: (1) Clifton Moss; (2) Four-Gates; (3) Id-Gate; (4) Stoney Lane; (5) Botany Bay; (6) Robin Hood; (7) St. George's; (8) Newton Road and (9) The Chemics. Our idea was to work on a number of plots so that the advice given would already be the result of experience from different trials of the same technique. Our methods cured erosion and corrected acidity. They grew grass with varying degrees of success from plot to plot, but none failed and so the methods used seemed worth trying anywhere. The author lived in Eccles at the time and kept a caravan in the Hindley or St. Helens neighbourhoods. Thus he was within easy reach of all the sites. An Arab riding mare provided the ideal means of travelling and surveying the ground.

In order to effect the generalization at which I aimed, it seemed necessary to design a simple and standard treatment and try it out on as many different sites as I could well manage. This came to nine sites over the first two years (Fig. 80). To the anti-erosion methods I added one refinement, namely to use a dressing of waste vegetation of some sort, it might be spoiled hay, it might be grass cuttings or garden rubbish, in order to give young seedlings shade and shelter and a little dampness and manure. One might call it making a micro-climate, on the very surface of the tip. So the method had to give stabilization against particle erosion, a microclimate, and a correction for acidity.

The method in all came down to: contour ridging (Figs. 81 and 82), sowing grass seeds with a very little general fertilizer, just to start them, strewing with a mulch of vegetable waste, chopped in so that it did not blow away or get washed off, pegging it down with sallow pegs (Fig. 83) and spreading lime. This was called 'frugal reclamation'.

The sallow pegs (Figs. 83–6) form another novel part of the treatment, which arose accidentally. The half-hay that I was using

81 The right riders. Riders can start revegetation by riding a pony straight up and down; sideways is risky and puts undue strain on the fetlock joints. (Photo. W. G. Mycock in the *Guardian*.)

82 Footprints. Because footprints aim to be level, the mare's on the right grew grass just as well as the boy's handiwork on the left.

83 Sallow pegs. Entrenching tools make good miniature mattocks for all the work on tips. As well as terracing we used them to hammer in eight inch pegs of sallow and poplar, until almost or completely out of the sight of meddling six year olds. Both actions shown are necessary to do the job quickly and well: start the peg then ram it in. (Photo. Joe Blossom.)

as mulch on one site was blowing on to the traffic on the trunk road A6 alongside it. I had to stop this and the only handy supply of pegs was a sallow bush. It was while hammering them in that I recollected that this kind of bush strikes very easily, and in fact about one in ten not only struck but survived the depredations of small children pulling them out during the Easter holidays. In later trials I always hammered them in until there was only an inch or less showing above the surface of the ground and the success rate usually varied from one in four to one in eight. However, one serious failure of sallows caused me to try burying the pegs just out of sight.

It will be remembered from Chapter 12 that the elephant grass in Africa charged downhill in one lump because it was not pegged down by the roots of bushes and trees. I had the same contingency in mind.

That reminds me that before I embarked on this campaign I had the helpful advice of a number of my ecologically-minded friends and colleagues. I suppose that Professor J. C. Mitcheson was the most expert of them, having himself tried and not always succeeded in tree planting on the tip of the well-known Birch Coppice Colliery, he being the third generation of managers of mines, and so familiar with the problem since childhood. Neither Mitcheson nor any other of my advisers will particularly want me to bother the reader with their names, but I should mention that

84 Growth from pegs *(left)*. A bush grown from the eight inch cutting visible, at Four Gates, Westhoughton, before being raised four and a half years later to be photographed at Salford University.
85 Slope of willows *(right)*. These were set a year before and left, neglected and trampled on, but they sprouted the new twigs shown and flourished. Survival varied according to the gradient of the slope: the result is best where the surface is too steep for the care-free traffic of children, of which there was not much on this site. (Photo. *Bolton Evening News.*)

86 Sallow patch. Among the trial plots of 1958–64, one included some steep bare shale at Clifton near Manchester. A proportion of pegs of sallow, set as mere cuttings, grew, and by 1971 provided good ground holding, looking permanent. Miss Jennifer Robinson, in the picture, had helped put in the eight inch pegs, as a child.

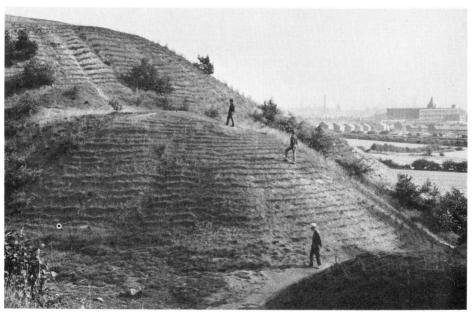

87 Terraces. These, made by a Friends' work camp at the Pretoria tip near Atherton, Lancs, imitate Brierley's terraces, shown in Fig. 79.

they included three farmers, one of whom had told me that I should probably need to use lime, another had also mentioned lime and the third waste hay as an extra supply of seed and encouragement to the new grass. None had experience of this work, but they seem to have had the insight.

The sites I used lay roughly within a triangle drawn from Salford to Westhoughton to St. Helens, which is some seventeen miles across (Fig. 80). One was on chemical wastes, mostly lime from a soda manufacture, one was on foundry slag and the remainder were on colliery shale. Under my methods, all nine sites grew grass, none failed entirely and with some after-care I am sure that all could be covered permanently, provided that sufficient sallows are used to cover the bare micro-precipices caused by the ridges, as shown in Brierley's diagram (Fig. 79), and provided that we solve the one sallow failure. Alternatively, wild white clover can cover the gaps.

Pilot scale

The nine-fold success with grass seemed quite sufficient to back a recommendation to clubs, schools and work-camps to use the techniques, and Friends' work-camps did so on two subsequent occasions, the one at Pretoria tip near Atherton (Fig. 87) being a conspicuous success with grass—the failure with sallows.

A tip which has had after-care and is a showpiece at the time of writing, is at Westleigh, done by the Westleigh County Secondary School gardening class and pleasing everybody in all possible ways. The school planned that when the grassing nears completion, they would plant it with small trees.

The Easter course: notes

Each Easter a practical course was held by the Biology Department at Salford, with two days devoted to reclaiming part of a slope of colliery shale by frugal reclamation, and the evenings occupied with lectures and discussions on the general subject of land reclamation. During those discussions questions were raised and answers given from the experience of the trial plots and subsequently. We may note these now.

On three sites the pH was tested after treatment and it had become 5·6–6. That seems to be sufficient for growing grass. Sallows will grow in lower pH and did in fact in one test strike at a pH of 3·5, which is very much on the acid side. On two sites powdered furnace slag was used successfully instead of lime. When a piece was flat enough to take farm implements the cost was under £20 per acre. I have since estimated the cost on slopes at less than

£150 per acre at prices of 1969. The wash from the tips, which is one of their most disagreeable features, was greatly reduced. On two sites, as well as testing bare shale, we also tested a wild sward that had invaded the tip and was struggling for survival. Merely liming it and sowing wild white clover made promising grazing. On special places such as conspicuous brows, or for small features, turfing is quite successful, as I proved on three sites. An uneven but vegetated site can be treated by lifting the turf and levelling, then adding lime and wild white clover. I did that on one only, even obliterating a picking pit in the process. Picking pits are holes about a yard and a half across and a yard deep where individual people have dug for coal. A surface dressing of ashes in one instance produced a notable addition to the benefits obtained from the standard treatment. Ground phosphate was added on one plot and seemed to be helpful, but was not always so.

As to the species of vegetation, the rye-grasses are the best to sow, as I tested on four sites; annual meadow grass was tested on three and is a useful pioneer. After these have taken a hold, fog and bent come in themselves. Doubtless if one continued to nourish the tip, one could keep the rye-grass, or alternatively cocksfoot, going, but leaving it to itself causes bents, fog and wavy hair grass to come in and take the place of the rye-grass. Among the clovers, alsike and red clovers are the best pioneers to sow, as I found on three sites, and wild clover shows up in about two years after sowing, as I found on six sites. Of the many herbs tried, chicory was by far the best, having shown up on three sites and survived. This is a very pleasant component.

Reference to plant succession in previous chapters will make it clear that the dressing of vegetable waste such as grass cuttings in the process of restoration is an attempt to substitute for the moss stage in natural regeneration, which by holding moisture allows grass seeds to germinate.

I think that the large seed reserve of the rye-grass enables it to start well, whereas bents and hair grass are needed to stay well. In neglected conditions, one needs a grass of low productivity, low demands in its root action, adaptable to unusual salts, such as the bents which have very little substance, or the hair grass which slowly builds a clump, adding a little each year. They seem greatly helped by the rye-grass starter.

Policy

It is hoped that schools in industrial areas will take up this work. The three in Lancashire or Yorkshire that have done so are pleased with the results. Many local education authorities are willing to sponsor teachers or senior scholars coming to special short courses like those held at Salford, who could then return and start the work. At the time of writing there is also a small fund in the hands of the Community Council of Lancashire from which the expenses involved in beginning this work on a new project can be met, whether in Lancashire or elsewhere. This is the fund left over from the Gulbenkian Foundation which financed the Friends' workcamps etc. following on the Development Commission trials.

In the whole picture of reclamation of derelict land, the frugal methods which have been mentioned so far are on a small scale, which is the intention of official policy, but they may form a great aggregate. In 1968, the Minister of Housing and Local Government sent a message to the Easter class giving it a general blessing and mentioning that grants would still be available for dealing with large sites but the intention is that the many small sites should be dealt with by the methods described, and it is hoped that schools, clubs and societies will use them. The students should, however, have some picture of the great projects that are in hand, and this is very conveniently given in John Oxenham's book *Reclaiming Derelict Land*. John Oxenham had many years experience in Housing and was a member of the Ministry of Agriculture Land Restoration Committee. There is a great variety of dereliction by various industries and considerable variety in the methods used to deal with each problem. In the description of the problem, and in the engineering methods used to restore the landscape, this book is thoroughly to be recommended, and on the more biological or ecological side it follows the accepted views of the time when it was written.

I have serious misgivings about the practice of restoring any site direct to agriculture. It is quite easy to grow a pasture or crop, using plenty of fertilizers, without really providing evidence of what that land will become over the years. Readers of Chapters 2 to 4 of this book will understand my anxieties. These have been fostered in my mind by conversations with Lancashire farmers who are my neighbours. They describe the so-called restored land as 'hungry' land, which needs a coating of muck every year and does not stand up in the way that traditional fertile land should. Agricultural advisers have confessed the difficulty that is caused by the strata of the subsoil being disturbed, as they put it, and I have wondered whether good agricultural land can be made over any subsoil that has not been under deciduous trees. Many tree roots go down into the subsoil, in most conditions, and when they rot they leave a drainage or seepage channel which the water will probably keep open for generations wherever the land demands it. In artificial fields there are no such channels.

The whole subject, to my mind, is quite open—I am anxious rather than sceptical—and if in the 1970s and 1980s there is a body of farmers who say that the restored land has proved satisfactory I shall be only too delighted.

I do remember that the first time I encountered this was in America where in 1949 our party saw the soil pushed off several acres of land in order to get out the brown coal underneath, and that same day the soil was pushed back again and was planted, not with an agricultural crop but with the false acacia, *Robinia*. *Robinia* is a very active tree belonging to the *Leguminosae*. It has a deep root and is a land improver. Since then, I have thought that the only method of restoring after open-cast or any other extraction work that an ecologist could guarantee, would be restoration to broad-leaved trees. In England, however, it was usually the law of the land that the sites must be restored to agriculture.

Looking at the whole question of the restoration of spoil land, it is surprising how much was done in the old days by colliery owners and great landowners. The work is now becoming increasingly important as the crowded inhabitants of cities seek fresh air and quieter scenery and I think that ecologists could find a useful role in advising on the reclamation of local eyesores.*

According to local circumstances and needs, there might or might not be room for 'frugal reclamation' methods on a larger scale, without earth-moulding. Hand labour is recognized as expensive, but so are the big machines used in moulding the landscape, and the question in any case has to be settled by accountancy. But the problem of roughing the slope has often been ignored, with consequent failure.

Such is the technical picture, but there are also social factors to be taken into account—in the next chapter.

---

* **Hazards**—Of course there may be hazards and some warning should be given. Aberfan, the Welsh village that was overwhelmed by a mountain of mud, should be mentioned to remind us that there may be lesser impoundments of water. To them we can add other hazards—shafts and other old workings, cavitation due to underground fires, etc.—against which a reclaimer will take precautions by keeping, as I always did, in close touch with the local authority's engineers. Once the hazard had built up, Aberfan was too big for them to tackle, but had their guidance been followed all along, it would never have occurred. An effective remedy for dangerous ground can only be biological, for farmers near industrial areas know that wire, however barbed, is ineffective against boys, but safety nets of branches and rootlets, e.g. of willow or poplar, should stop serious accidents. This question is discussed in Chapter 15.

Reclamation of spoil lands/II/social   15

Introducing children
The results described in the previous chapter were obtained by a course of action that really began at Lowestoft with daffodils. Each year the daffodils grew until the buds were showing yellow and then we never saw them again. Somebody picked them before they opened. One spring I mentioned this to the local policeman and we made a plan. He would keep an eye on the road just outside my place, and I would put up a notice saying, 'Beware of the Bull' in a place where trespassers could not miss it as they came through the hedge. Within two days of my putting up the notice a little deputation of children knocked at my door and their spokesman said: 'Please Mr. Graham, may we see your bull?'

I ought to have known you can never get the better of children. The only result was a rather embarrassing conversation about the difference between the bull I did not, after all, buy and the bullocks they could see that I had bought. Finally, I said to them: 'Are you interested in daffodils?'

They said they were, so I asked them if they would like to work in my garden and then I would give them some daffodils. It became customary that we worked that way, and from that year onwards I did not lose my daffodils. In addition to rewarding them with daffodils, I had them write their names and addresses in a book, which seemed to interest them, although I did not at the time tell them what I had in mind. Eventually, it turned into a children's garden party.

It appeared that the children came and would work quite well, chiefly for the sake of having their names in the book. The last task that the Lowestoft children did for me was to collect grass seeds for greening over the tips in Lancashire, which they seemed to think was a good idea.

When I came to Lancashire, therefore, it was fairly obvious that I should get all the help I needed if somehow I could get the children interested (Fig. 88). In fact, the children always came as soon as they saw me working on a tip, especially if I had a horse with me. Again they found it important to write their names,

88 Young energy. Admittedly my Arab riding mare, a beauty from a royal stable, meant that I had an easy introduction anywhere, especially to children. Nevertheless, I thought that their basic motive for helping me so often and so well, was the outlet for their energy provided by projects approved by them, and in which they would not be ordered about. There were always tools in the caravan for them to use.

addresses and ages in my note-book and the rewards were two: when I took snapshots, I sent them prints; and at Christmas each received a card. In 1960 there were seventy-three children for Christmas cards and the number rose to about a hundred and thirty in the three or four years that I was active on the tips.

I travelled and worked from a horse caravan, which happened to be a gypsy type, and this proved to be an additional attraction for the children, who loved to come and sit in it. Sometimes when it rained I had as many as sixteen children packed into the van and an adult visitor was astonished at the spectacle that met him when he opened the door.

The whole project proved popular. The children not only came but they brought gifts of food and tools from their homes, for which they would take no money.

People often ask how I managed to organize the children to help me: I did not organize them at all, they were there on the tips playing and when I came they joined in with what I was doing.

Recreation
Independently of me, they were using the tips in ways that people who merely pass by do not know anything about. It is a very interesting experience to hurl a stone over the wall guarding a shaft and to hear it splash into the water an incredibly long time after. This particular recreation does not please the adult world and may indeed interfere with some of the arrangements of the National Coal Board, which owns the shafts and is responsible for keeping them safe. At one tip the shafts were capped with concrete and provided with ventilating pipes so as to let gas escape. It was a gassy tip which had been closed after a serious explosion in the early part of the century. The concrete cap was fenced with iron spikes eight feet high, but the children had found it possible to burrow under it, like rabbits under an inefficient rabbit fence, and they enjoyed hearing pebbles splash into the water when they dropped them through the ventilating pipe. By some mischance one of the stones stuck in the pipe and then more stones were dropped in and the escape for the gas was sealed off. As a result there was another explosion and the concrete cap was broken.

For a short time the Community Council and I took responsibility for the safety of the shafts and we carefully built up all the holes so that they were secure, but afterwards they were opened up again and it was possible to put quite large objects such as unwanted corpses down through the holes into those shafts. Latterly, however, trees have grown up inside the fence sufficiently to cover up the whole problem—and I think solve it.

In one way and another, then, the shafts can be hazards. Many of the tips are built over ancient and long-forgotten shafts, which were sealed off by putting timbers across a few feet down and then filling in on top. When the timbers rot, whole houses may disappear down coal shafts—lost in the carelessness of bygone centuries.

There are perhaps two satisfactory ways of dealing with shafts. When they are known and open they can be filled up, but as far as the unknown and long-buried shafts are concerned, it seems to me to be essential to get trees growing right over their suspected location, because it is unlikely that a whole tree would go clean down a shaft taking children or anybody else with it. It would surely jam across somewhere. Positions are often known approximately.

I found that the tips provide many other recreations of a more innocent kind; one of the favourites is what can be described as Alpine bicycling (Fig. 89a) which gives the larger boys immense joy, going up and down the hillocks of what is sometimes called lunar landscape. Another is tobogganing on sheets of metal (Fig. 89*b*) which provide a good deal of fun. Many tips have the old reservoir, or 'lodge', in which there are fish and these are fished assiduously, some indeed being rented by angling clubs. On others there is shooting for hares or at least rats. It is surprising on what remote parts of barren coal tips you may find a hare lying up. Hundreds of dogs are exercised on the tips in various parts of south Lancashire, where there are all too few open spaces where a man

89 *Their rucks.* Thus (for example) were the familiar rucks used by children for recreation: (*a*) (*above left*) tobogganing on sheets of iron; (*b*) (*above right*) alpine bicycling; (*c*) (*below left*) nature walks and horse riding; and (*d*) (*below right*) a lively run.

and his dog can get the air. There is bathing; I have seen archery.

In St. Helens a man gave me quite a lecture on how good the air was up on top of a high tip. On another, a lady picnicking with her children said: 'You know, the air up here is the same as at Blackpool.'

The tips do in fact very often hide one depressing part of a town from another.

The best recreation of all for children on tips is simply running up and down (Fig. 89d). A steep slope excites them and they rush down it with every sign of enjoyment. If there is a hazard at the bottom, such as a dirty stream into which they must not fall, so much the better it seems. They can get into little canyons and valleys in the tips and play at 'house' or camping. They can indulge in mountain warfare and play cowboys and Indians. Riders come with their ponies. All this amounts to a great deal in our England of which the agricultural parts are too much taken up by the Enclosure Acts; moorlands forbidden for waterworks or game; and every accessible open amenity overcrowded. Furthermore, these play areas are often on the door-step of houses in grey areas. It is as if they were living right alongside some fine unexplored adventure country; such as we could not afford to provide, even if we knew how.

When explaining this once to a significant visitor, representing the Gulbenkian Foundation which was backing the work, it happened that we stood on a ridge of true slag east of Wigan, and from a school perhaps three-quarters of a mile away there suddenly debouched figures in running outfits. The boys came running by us across the field of slag and other waste land and disappeared northwards, except for a small party which broke off from the others to swim in one of the lodges, rejoining when the cross-country run came back. We agreed that there was here surely a unique facility for the school in that at their very back entrance they could organize cross-country runs, to say nothing of swimming, without getting permission from anybody, entirely clear of traffic and in air that seems clear and healthy enough now, although when I was a boy Wigan's air was nothing but dust and smoke and the furious sound of incandescent slag exploding as it came rolling down the tip.

Frustrations and vandalism

During the four years that I worked on the tips, there were changes. Gradually the district was 'improved': a movable pontoon that had enabled me to cross a canal was padlocked to one side as the canal was no longer in use. This meant that I had to go two miles round with my horse, whereas before I had got across with the co-operation of the numerous children who voyaged on the pontoon, moving it from one side of the canal to the other. Another favourite way of mine led across a farmer's fields by a footpath on which he put padlocked gates and stiles. I offered to pay for permission to cross by the footpath but he said that no matter how much money I offered him he did not wish to let me and my horse go across. I saw his point. A favourite place for picnicking, by a disused canal where there were some interesting exotic flora, was reclaimed, as it is called, and made into a dull field through which I could not pass. The shaft to which I have referred as a dramatic stone-dropping point was filled in. I am bound to wonder what those youths do instead. It was no child's play to loft the stones over the shaft's high wall.

In the same period, I began to hear more and more about vandalism. More and more public lavatories were closed, to the discomfort of the Queen's lieges, because people had ripped out the copper or brass piping. Finally one of the bathing places on spoil land, where I had watched young people sporting during an early fine summer—and which was the only recreation place of any kind in a particularly drab outlying village—was embodied in another dull field, and the undressing 'rocks' of clinker were broken up and scattered.

There was in fact a great move to tidy up the appearance of the industrial landscape as seen by the passing motorist, and the Civic Trust was founded with the intention of making everything more presentable. I do not object to that in the least, in fact I have all along supported the Civic Trust and have friendly relations with them, but it is quite obvious that their emphasis is all based on the values of middle-class adults passing through, instead of being on the values of working-class children who live there. Waste pieces on which twenty or thirty boys played informal games were tidied into gardens—sure enough, but the boys then had to play ball games on side streets.

The work that I was doing was noted for suffering from vandalism. I had one very considerable experience of that, because the Friends' Work-campers who were helping me would not believe that it was necessary to do literally and precisely what I said. They put in about three thousand cuttings of poplar on some waste land at Bolton and instead of cutting them off flush with the ground as is my method, they left them as tall sprays and said didn't they look lovely. After a mild protest, I had to leave it at that, because there are limits to what one can do with volunteers and their leaders. Three days later every single one of the cuttings was out on the ground and dried up. I made enquiries of the local children, who said that it had been done by a little party of six and seven year olds. I asked why they had done it and whether they themselves would have done the same at that age. These children were between nine and twelve years old. They said that they would have done it—'just for something to do'.

My conclusion from my experience is that the supply of playing fields, or even of swimming baths, will not meet this particular requirement of young energy. They need to do something which they have thought up themselves, and at once. That, after all, is what adults wish to do, and that seems to me a fundamental point in human ecology and one which is not yet provided for by any great organization or policy.

Throughout the work there has never been the slightest difficulty in obtaining all the child and adolescent volunteer help required. The reason why the idea has not caught on very much is that there

90 Playsward on colliery shale. In a trial carried out with the County Planning Officer a soft and attractive turf was made in one year without carting on soil but by growing and then chopping in rape, red clover and Italian rye grass, well-grown on the shale after a light dressing of sewage sludge.

are very few adults able or willing to provide the continuity, doing for their local eyesores what I did for the nine plots. I made no bargain with the young people; I just went there and started to work and sometimes they came in sufficient numbers and sometimes they did not. There was a provision of tools and materials if they did come to use them. If anyone says that that is a very unsatisfactory way of working, my answer is that until the invention of steam power men were able to do great things with the use of intermittent power from windmills and watermills. On this sort of work, if it is to give them pleasure, the children must work for themselves.

So the nett result of attempting social landscapers' clubs is confined to one or two sites of Friends' Work-camps. As mentioned in Chapter 14, three schools have done very well and Salford University took over the courses, which are firmly pointed at schools.

The vandalism on Rivington Pike, mentioned in Chapter 12,

91 Scars. Industry is said to have scarred England. Here, however, it is because of the colliery site that this wood has been allowed to grow *(left)* and provide a place for picnics and other forms of relaxation. Without the presence of the colliery, it would still be a rather dull field as its neighbours are. These alder of quality *(right)* (mainly natural regeneration) growing on bare colliery shale are about 80 years old.

may also be recalled as a conspicuous example of the problem of young energy. Once such blemishes are seen to be due to young people's energy, they can hardly be considered only as a matter of regret. It would not be healthy if young people were not so energetic. That I think is a fairly safe conclusion in human ecology. The problem is how to use that energy gracefully in a way that can yet be at the choice and will of the potential vandals, as it must be if it is to work.

16 *Ecology in progress*

The theory of determinants
In Chapter I we saw how the search for determinants arose as a teaching device. It was a framework on which to hang facts and thoughts about common eco-systems and common ecological processes.

Now it will serve for recapitulation.

Thus, we said that keeping herbage short was the determinant of a pasture or meadow, which turned into a wood if left to grow.

Chapters 2 and 3 showed the parts played by many members of the animal kingdom in re-making the humus component of soil from vegetable remains. Thus the animals serve mankind in renewing soil fertility. This suggested determinant was fully demonstrated in Chapter 4, where we learned that probably some millions of English acres are spoilt for growing wheat because farmers no longer practise animal husbandry. Thus those myriads of soil organisms that can feed on either humus or crops have no alternative but to attack the crop. Loss of insulation, aeration and stability are equally important. In the 1960s we began to see English Downs with no sheep on them. In Chapter 12 it became obvious that the animal kingdom made the first soil by consuming rootless plant forms, after which there could be something for roots to inhabit and exploit. The laying waste of the fertility of the great continent of North America was due to greedy farming.

Not all the chapters gave us definite evidence from macro-ecology to show that the determinant named has been diagnosed correctly. In many cases the choice of determinant finishes by being no more than a plausible theory. In Chapter 5, mud was first proposed as the determinant of life in a pond but was abandoned in favour of clean water supply; current was for a stream, thermocline for a lake, particle size in an estuary, exposure for a rocky shore, fishing for nearby seas, and the rotation of the earth for the ocean. In the case of the ocean the variation with latitude in the warmth of the sun could be joined with the Coriolis force. In forestry we thought the determinant to be the volume of standing timber and experience has shown that success often depends upon attending to that.

For woodland in Chapter 7 we learnt to look first at the state of the canopy. Bird migrations seemed to be determined by seasonal fluctuations of the birds' food. Fish migration seems likely to be compensation for the drift of the feeble progeny in the plankton or down streams, but with possibilities of the same nourishment determinant as for birds.

In Chapter 9 we thought that there was a strong case for saying that the determinant of a healthy population of prey is the presence of predators. The territorial behaviour of the prey in putting the members of lower status in the most vulnerable positions renders the action of the predator markedly therapeutic.

In the tropical great lake in Chapter 10, the south east trade wind was found to be the prime determinant for most of its ecology. Chapter 11 was mostly concerned with techniques of numbers, but in doing so it showed that moderation of predation, in this instance by man, is the determinant of the best exploitation: the optimum catch.

In anti-erosion problems of Chapter 12 we could cite the moderation as in Chapter 11, thus maintaining the humus content as in Chapters 2 to 4, and using a proper balance of trees with animal husbandry, which calls on the knowledge referred to in Chapters 2 to 4 and 6 to 7.

In Chapter 13 we saw that the undisciplined rivers are determined by heedless grazing. In Chapter 14, we found that erosion often prevents vegetation of industrial spoil, in Chapter 15 we suspected that nothing more sinister than youthful energy is the determinant of vandalism. Those examples show that one may, with more or less degree of probability, assign to the complex little worlds of eco-systems, one or two factors which commonly dominate each kind.

The labelling should not be carried too far. From their nature, as we have just surveyed them, we can see that there is the following risk in the theory of determinants.

The use of the word determinant may suggest pre-determination or rigidity, which is very far from being characteristic of eco systems. Ecology contains many unexpected outbreaks of animals and plants: starlings invading city buildings for roosts; the hares' jamboree on Belfast airport (Fig. 92); kestrels over the trunk roads; coypu clearing waterways and ditches as referred to in Chapter 1, and water-courses becoming clogged with unwanted vegetation as a result of excessive run-off of nitrates from agricultural soil.

230   A natural ecology

92   Ancient lagomorph adapts. According to *The Times*, 25 November 1966, more than fifty hares could often be seen taking their ease on the grass between the runways at Aldergrove Airport, Belfast, protected there from dog and gun. Thus, uninstructed, they gained from human activity as animals often do. (Photo. *The Times*.)

In the last case and in some others the outbreaks are explicable after the event and so the determinant inferred.

Apart from their teaching value, there is some degree of sense in having these determinants in mind: partly because it should prevent important factors being overlooked and partly because eco-systems which are apparently incomprehensible may after all be sorted out and eventually be found to be controllable. Undoubtedly in eco-systems determinants exist, as in other departments of life, but quite often we do not know them, and there are plenty of variants. Forestry for example, to which we have assigned the volume of standing timber as the determinant, may be manifested in some woods as an expression of recreative value, whether in cover or game or anything else. In this case the management of the forest would be different, as it would also be if it was being used substantially for the grazing of cattle. Evidence from history and from pilot-scale trials is always required before ecologists can feel really confident in giving advice. One must feel one's way among the variants.

That was as far as I went in the course.

A common ecology? Advice
One may ask whether the search in this book has led to any ecology applicable to all eco-systems. The chapters seem to give the following, which can be expressed as strong advice:

(1) to discover the determinants;
(2) to test for variants by historical evidence or pilot-scale;
(3) to take only the natural interest of a stock, itself the natural capital conserved to the level of best average yield, the 'optimum' including consideration of all circumstances;
(4) to preserve the aerobic situation in air, water and soil and to monitor acidity;
(5) to allow sufficient predators and scavengers, not overlooking migrants; and
(6) to restrict all poisons.

Scholasticism: logic on too few factors
I continue the book a little, because my recent experience taught me something about teaching that may, if correctable, help with the human predicament.

Teachers work the personal magic that helps to fashion some very fine people—I meet them every day that I go into Salford University. But some people might say that magic is always dangerous, and I admit that there is danger here.

Peter said: 'Mr. Graham, I don't see how there could be an examination on this subject. You take us to it, we see it; and that's that.'

He had a real point, but it was in the orders to have an examination and I provided questions—one compulsory question and some essay subjects. The compulsory question ran in some variant of the following: 'What determinants would you expect to find in the following eco-systems: the farm; a stream; [etc. etc.]? Discuss the action of determinants in two of the eco-systems mentioned.'

There would be nothing to prevent a pupil answering the first part of the question by mere memorizing of his lecture notes—as if it were a question of dates of kings. Let us pursue that possibility for a moment.

Let us suppose that this type of ecology became a popular subject, taught in many universities and in all schools—as one hopes it will be. There would then be a danger in that teachers would find it convenient to have available a table of determinants, like the imaginary ones that my pupils had in their heads when they went into the examination room. Suppose that gradually the phrase 'expect to find', from my text, lost its emphasis. Then it might easily come about that the theory of determinants would become a piece of scholasticism, tidy, easy to teach, easy to mark, but suggesting far too few factors for practical life. There's the rub, in my argument; there the world goes wrong: the neat and easy to teach. If that happened in this instance, I should count any success of this book to be turned into failure. Those pupils need have none of the inkling into landscape that the course was designed to give. I should picture graduates being turned out as 'know-alls' in ecology and as such derided by experienced men, disliked at sight, and on instinct, by common men. Those graduates, according to the custom in other subjects, might go immediately into teaching; or they might make a living by writing; or they might make speeches; or perhaps they would go into offices and manage other men. Unfortunately their whole training has been to make short and clear statements based necessarily on too few factors. The like runs through all academic subjects that I have encountered and so could bring it about that the world is ruled by parrotry.

Scholasticism sins by overstepping the proper bounds of theory. In geometry, a line has no thickness—none of the logic would work if it had; but in map-making the first step is to determine the thickness of your line. In the theory of levers the sums assume them weightless, but there would be some catastrophes if the engineers did so. In life, logarithmic growth, calculus, and probability are realities that can be drawn as pictures and are easily comprehensible; but in the past they have been hidden in algebraic juggling so that a great many people are prevented from acquiring the quantitative inkling that all men need into life processes.

The question arises whether such ill effects are inevitable in true education. Let us take Stapledon's criticism in his book *Human Ecology* which includes a considerable discussion of what an educated man should be, and how he should learn.

Stapledon pointed out that there are three methods of learning; one by listening, as in a class; another by watching, as in an apprenticeship, and a third by conditioning or practice, with the mistakes, modifications and improvements that occur.

I would add that the first method necessarily mentions too few factors. The second shows a skilled man's selection of more factors; and in the third the learner gradually gets the feel of the subject, so that many factors—those that the pupil himself needs—go into his final mastery of it.

Let us not deny that theory is essential, to give scope and versatility, as in the examples of the 45° triangle in Chapter 6. But theory is probably over-valued in policy for 'Education'. Stapledon evidently held that concentrating all a man's learning years on this first method, by listening, would be a serious fault, with far-reaching consequences. It could do harm by habituating the pupil to taking account of the main factors only, as they were seen by the teacher. Unless he imparts them clearly, the pupil, or newspaper reader, or televiewer, will be in a muddle; but if given theory only, the recipient is pushed towards 'single factor analysis' as Lewis Mumford calls it in *The Myth of the Machine*. He convincingly shows that single factor analysis is responsible for many of the suicidal tendencies of the 20th century.

Clearing the decks

Teachers can put their own world right—indeed in the 1970s many teachers are in an open-minded mood. The younger people, however, seem deeply unsettled, across all the world. To account for their distress, there are truly enough dangers that threaten their future lives, more than can be banished by music and universal love.

It might help some of them if they could sort out the dangers by applying the methods of thoughtful natural history.

Some examples will show the sorting.

Thus: the bombs? Yes: sufficient exist to over-kill the world. Yes: it is general experience that humans err. So: the end may come any day, starting with tomorrow.

The veil of dust around the world makes haze where there should be the warmth of clear sunlight, and so threatens an Ice Age. Is the dust volcanic or man-made? The Krakatoa volcanic explosion loaded the world skies with dust for a decade or two during the last century. The dust enriched the colours of the sunset skies everywhere. Then it disappeared. Will this?

The milk of American women contains higher than normal DDT, which persists and piles up in the liver. Does this matter? You should be able to answer. How can the use of poisons be restricted?

After asking a few more of such questions you should be able to select the ones demanding immediate international action, after cool enquiry. It is no good losing your head today even if it may be lost for ever tomorrow. It may not.

Bibliography

Albrecht, W. A.: 'Physical, chemical and biochemical changes in the soil community', in Wenner-Gren Symposium (see below), 1956.
Allaby, M.: *The Eco-Activists*, 1971.
Apstein, C.: 'Die Bestimmung des Alters pelagisch lebender Fischeier', Mitt. der *Deutsch Seefisch*, vol. xxv, no. 12, 1909.
Arago, F.: 'Annuaire du Bureau des Long', 1844. Translated among Arago's *Biographies of Distinguished Men*, 1857.
Arrhenius, O.: *Kalkfrage, Bodenreaction und Pflanzenwachstum*, 1926.
Baerends, G. P. and Baerends-Van Roon: 'Ethology of cichlid fishes', in *Behaviour*, supp. 1, 1950.
Balfour, Lady E.: *The Living Soil*, 1949.
Beresford, T.: 'How the farmer sees it', in *The Countryman*, Autumn 1967.
Bertalanffy, L. von: 'The theory of open systems in physics and biology', in *Science*, vol. 3, no. 23, 1950.
Bertin, L.: *Eels*, 1956.
Beverton, R. J. H. and Holt, S. J.: *On the Dynamics of Exploited Fish Populations*, 1956.
Billington, F. H.: *Compost*, 1940.
Black, J. D.: *Biological Resources*, 1968.
Brenchley, G. H.: See issues of *Journal of the Farmers' Club*, 1968–1970, for the full debate on soil fertility.
Brierley, J. K.: 'Some preliminary observations on the ecology of pit heaps', in *Journal of Ecology*, vol. 44, 1956.
Brown, H.: See issues of *Journal of the Farmers' Club*, 1968–70, for the full debate on soil fertility.
Buxton, P.: *Animal Life in Deserts*, 1923.
Carrighar, S.: *Wild Heritage*, 1965.
Carruthers, J. N.: 'New drift bottles', in *Conseil Int. J.*, vol. III, no. 2, 1928.
Cave, T.: See issues of *Journal of the Farmers' Club*, 1968–70, for the full debate on soil fertility.

Chrystal, G.: 'Seiches and other oscillations of lake-surfaces, observed by the Scottish Lake Survey', in *Bathymetrical Survey of Scottish Freshwater Lochs*, vol. 1, 1910.
Clarke, G. L.: *Elements of Ecology*, 1954.
— *Introduction to Ecology*, 1971.
Cooper, C. F. 'The Ecology of Fire' in *Sci. American*, April, 1961.
Crowcroft, P.: *Mice all over*, 1966.
Darling, F. Frazer: *A Herd of Red Deer*, 1937.
Debenham, F.: *Map Making*, 1936.
Dobbs, C. M.: 'Fishing in the Kavirondo Gulf, Lake Victoria', in *Journal of E. Africa and Uganda Nat. Hist. Soc.*, no. 30, 1928.
Dorst, J.: *The Migration of Birds*, 1962.
Edwards, V. C. Wynne: *Animal Dispersion in Relation to Social Behaviour*, 1962.
Elliot, R. H.: *Agricultural Changes and Laying Down Land to Grass*, 3rd. ed., 1905.
— *The Clifton Park System of Farming*, 4th ed., 1908. Last published 1943.
Ellis, E. A.: *The Broads*, 1965.
Elton, C. S.: *Animal Ecology*, 1927.
— *Animal Ecology and Evolution*, 1930.
— *Ecology of Animals*, 1933.
— *Exploring the Animal World*, 1933.
— *Voles, Mice and Lemmings*, 1942.
— *Ecology of Invasions by Animals and Plants*, 1958.
— *The Pattern of Animal Communities*, 1966.
Fell, H.: See issues of *Journal of the Farmer's Club*, 1968–70, for the full debate on soil fertility.
Forsyth, S.: 'A river runneth', in *The Countryman*, Winter, 1967.
Frankland, J. C., Ovington, J. D. and Macrae, C.: 'Spacial and seasonal variations in soil, litter and ground vegetation in some Lake District woodlands; in *Journal of Ecology*, vol. 51, pp. 97–112, 1963.
Furneaux, B. S.: See issues of *Journal of Farmer's Club*, 1968–70, for full debate on soil fertility.
Galton, Sir F.: *Natural Inheritance*, 1889.
—*Hereditary Genius*, 1869.
—*Human Faculty*, 1883.
Gardiner, R.: 'Forestry and husbandry', in *The Natural Order*, ed. H. Massingham, 1945.
Graham, M.: *The Victoria Nyanza*, 1929.
— 'Modern theory of exploiting a fishery', in *Conseil Int. pour l'exploration de la Mer*, 1935.

— 'Tilapia esculenta', in *Annals and Magazine of Natural History*, 1928.
— *Soil and Sense*, 1941.
— *Fish Gate*, 1943.
— 'Science and British Fisheries' in *Sea Fisheries*, 1956, ed. M. Graham.
— 'Crowds and the like in vertebrates', in *Human Relations*, vol. 17, no. 4, 1964.
— 'Swift rivers', in *The Countryman*, Winter 1968.
— 'Young Energy', in *Town and Country Planning* Oct. 1968.
— and Harding, J. P.: 'Some observations of the Hydrology and Plankton of the North Sea and English Channel', in *Journal of the Marine Biological Assn. of the U.K.* Vol. xxiii, pp. 201–6, November 1938.
Gulland, J. A.: *Estimation of Growth and Mortality in Commercial Fish Populations*, 1955.
Haeckel, E. H.: *General Morphology*, 1866.
— *History of Creation*, 1892 (English trans.).
Hall, Sir D.: *A Pilgrimage of British Farming*, 1913.
Hall, I. G.: 'The ecology of disused pit heaps in England', in *Journal of Ecology*, vol. 45, 1957.
Hanson, C. O.: *Forestry for Woodmen*, 3rd. ed., 1934.
Hardy, Sir A. C.: *The Living Stream*, 1965.
— *Great Waters*, 1967.
— Lebour, M. and Savage R. E.: *The Open Sea*, 1959.
Haughley Research Reports 1938–62, ed. R. Waller, (Soil Assoc.).
Heape, A. S.: *Emigration, Migration and Nomadism*, 1931.
Hediger, H. P.: 'The evolution of territorial behaviour', in *Social Life of Early Man*, ed. S. L. Washburn, 1962.
Hensen, V.: 'Über die Bestimmung des Planktons oder des im Meere treibenden Materials an Pflanzen und tieren', in *Wiss. Meeresuntersuch*, vol. 1, 1887. (There is an account in English in *Sea Fisheries*, ed. M. Graham.)
Hjort, J., Jahn, G., and Otterstadt, P.: 'The optimum catch', in *Hvalrådets Skrifter*. 1936.
Huntsman, A. G.: 'Effect of predation on the salmon fry', in *Journal of the Fishery Research Board*, 1941.
Huxley, T. H.: *International Fisheries Exhibition, 1883*. The fisheries exhibition literature, vol. iv: Conferences, pt. 1; Inaugural address by Professor Huxley. 1884.
Jamieson, A. and Jonsson, J.: 'The Greenland component of spawning cod at Iceland', in *Conseil Int. pour l'exploration de la Mer, extrait du rapports et procès verbaux*, vol. 161, 1971.

Jones, N. S.: 'Observations and experiments on the biology of *Patella Vulgate* at Port St. Mary, Isle of Man', in *Liverpool Bibl. Soc.*, vol. 56, 1948.
Jones, R. H.: *Migration of Fish*, 1968.
Kershaw, W. E., Leytham, G. W. H. and Dickerson, G.: 'Effect of schistosomiasis on animal intelligence', in *Annals of Tropical Medicine and Parasitology*, vol. 53, no. 4, 1959.
Kinship Group: See under H. Massingham, ed.
Kirkman, F.: *The British Bird Book*, 1911.
Kluvers, H. N.: *The Population Ecology of the Great Tit*, 1951.
Kramer, G.: 'Experiments on bird orientation', in *Ibis*, vol. 94, 1952.
— 'Experiments on bird orientation and their interpretation', in *Ibis*, vol. 99, 1957.
— 'Recent experiments on bird orientation', in *Ibis*, vol. 101, 1959.
Kropotkin, P. A.: *Fields, Factories and Workshops*, 1898.
Laplace, M. de: 'Sur les naissances, les mariages et les morts a Paris, depuis 1771 jusqu'en 1784; et dans toute l'étendue de la France, pendant les années 1781 et 1782', in *Histoire de l'Academie Royale des Sciences, avec les memoires de Mathématique et de Physique, pour la même année*, 1783.
Laplace, Du Sejour and De Condorcet: 'Suite de l'essai pour connoitre la population du Royaume, et le nombre de ses habitans, en adaptant aux villes, bourgs et villages, portes sur chacune des cartes de M. de Cassini, l'annee commune des naissances et en la multipliant par 26', in *Histoire de l'Academie Royale des Sciences, avec les Memoires de Mathématique et de Physique, pour la même année*, 1784 and 1788.
Le Bon, G.: *The Crowd*, 1896.
Lebour, M. V.: 'The food of young fish', in *Journal of the Marine Biol. Assoc.*, vol. xii, 1919.
Leopold, A. S.: 'Ecological aspects of deer production on forest lands', in *Proceedings of the U.N. Scientific Conference on the Conservation and Utilization of Resources*, 1949 and vol. vii: *Wild life and fish resources*, 1951.
Liebig, J. von: 'Chemistry in its application to agriculture and physiology', Report to the British Assoc. for the Advancement of Science, 1840.
Lincoln, F. C.: *Migration of Birds*, 1952.
Macan, T. T. and Worthington, E. B.: *Life in Lakes and Rivers*, 1968.
Maio, J. J.: 'Predatory Fungi' in *Sci. American*, vol. 199, no. 1, July 1958.

Malthus, T. R.: *An Essay on the Principle of Population*, 1890.
Manly, B. J. and Parr, M. J.: 'A new method of estimating population size, survivorship and birth rate from capture-recapture data', in *Transactions of the Society for British Entomology*, Dec. 1968.
Marais, E. N.: *My Friends the Baboons*, 2nd. ed., 1947.
Massingham, H. (ed.): *The Natural Order: Essays in the Return to Husbandry*, 1945.
Meek, A.: *The Migrations of Fish*, 1916.
Merricks, J.: See issues of *Journal of the Farmers' Club*, 1968-70, for the full debate on soil fertility.
Miller, E. C.: *Plant Physiology*, 1938.
Morand: 'Recapitulation des baptemes, mariages, mortuaires et enfants-trouves de la ville et faubourgs de Paris, depuis l'année 1709, jusques et compris l'année 1770; precedee de quelques remarques generales sur ce tableau', in *Histoire de l'Academie Royale des Sciences, avec les Memoires de Mathématique et de Physique, pour la même année, 1779*.
— 'Memoire sur la population de Paris, et sur celle des provinces de la France, avec des recherches qui etablissent l'accroissement de la population de la capitale et du reste du royaume. Depuis le commencement du siecle', in *Histoire de l'Academie Royale des Sciences, avec les Memoires de Mathématique et de Physique, pour la même année, 1779*.
Morrow, G.: *Quoth the Raven*, 1919.
Mumford, L.: *The Myth of the Machine*, 1968.
Murray, Sir J.: *The Voyage of HMS 'Challenger': a Summary of Results*, 1895. (nb. The Challenger Office, Edinburgh has numerous other references to wind induced currents.)
Neal, E. G.: *Woodland Ecology*, 1958.
Norman, J. R.: *History of Fishes*, 1931.
Oxenham, J.: *Reclaiming Derelict Land*, 1966.
Oyler, P.: *The Generous Earth*, 1950.
— *Sons of the Generous Earth*, 1963.
Parr, M. J.: 'A population study of a colony of imaginal *Ischnura elegans* (van der Linden) (Odonata: Coenagriidae) at Dale, Pembrokeshire', in *Field Studies*, vol. 2, 1965.
Pearsall, W. H.: 'Soil sourness and soil acidity', in *Journal of Ecology*, vol. 14, 1926.
— 'Soil complex in relation to plant communities', in *Journal of Ecology*, vol. 26, 1938.
Percy, R.: See issues of *Journal of the Farmers' Club*, 1968-70, for full debate on soil fertility.

Petersen, C. G. J.: *On the Decrease of Flat-Fish Fisheries: Report of Danish Biological Station*, vol. iv, 1894.
Popham, E. J.: *Some Aspects of Life, in Fresh Water*, 2nd. ed., 1961.
Popham, E. J.: *Some Aspects of Life in Fresh Water*, 2nd. ed. vitamin D milk on dento-facial structures of experimental animals', in *American Journal of Orthodontics and Oral Surgery*, vol. xxxii, 1946.
Richardson, J. A.: See *Planning Outlook*, 1967 (University of Durham).
Robinson, D. N.: 'Soil erosion by wind in Lincolnshire, March 1968', in *East Midland Geographer*, vol. 4, pt. 6, no. 30, Dec. 1968.
Ruskin, J.: *Modern Painters*, 1888.
Russell, E. S.: *The Overfishing Problem*, 1942.
Russell, E. W.: *Soil Conditions and Plant Growth*, 9th ed., 1968.
Salisbury, E. J.: See *New Phytology*, vol. 14, 1915.
—See *Quarterly Journal of Royal Met. Soc. Vol. 65*, 1939.
Sauer, C. O.: 'Sedentary and mobile bents in early societies', in *Social Life of Early Man*, ed. S. L. Washburn, 1962.
Savage, R. E.: 'The relation between the feeding of herring and the plankton', in *Fisheries Investigation*, ser. 2, vol. xii, no. 3, 1931.
Savory, C. A. R.: 'Crisis in Rhodesia', in *Oryx*, the Journal of the Fauna Preservation Society, vol. 10, May 1969.
Schmidt, J.: 'The breeding places of the eel', in *Phil. Trans.* ser. B, vol. ccxi, 1922.
Sharp: *Insects*.
Simpson, A. C.: 'The fecundity of the plaice', in *Fisheries Investigation*, ser. 2, vol. xvii, no. 5, 1951.
Southern, H. N.: 'The ecology of population dynamics of the wild rabbit', in *Annual Applied Biology*, vol. 27, 1940.
Southward, A. J.: *Life on the Seashore*, 1965.
Stapledon, Sir R. G.: *Hill Lands of Britain*, 1937.
— *Human Ecology*, 1964.
Steinbeck, J.: *The Grapes of Wrath*, 1939.
Straelen, Mme. van: *Victor van Straelen: a Memoir*, 1964.
Taaning, A. V.: See *Conseil Int. pour l'exploration de la mer*, vol. xii, no. 1, 1937.
— See Graham, M. (ed.) *Sea fisheries*.
Thomas, W. L. jr.: See Wenner-Gren Symposium.
Thompson, W. F.: 'Conservation of the Pacific Halibut', in *Smithsonian Report for 1935*, 1936.

Thompson, W. F. and Bell, F. H.: 'Biological statistics of the Pacific halibut fishery', 1934.
Tills, D., Mourant, A. E. and Jamieson, A.: 'Red-cell enzyme variants of Icelandic and North Sea cod *(Gadus Morhua)*' in *Conseil Int. pour l'exploration de la Mer, extrait du rapports et proces verbaux*, vol. 161, 1971.
Waller, R.: See *Haughley Research Reports.*
Watson, G. G.: *Fun with Ecology*, 1967.
Wenner-Gren Symposium. W. L. Thomas jr. (ed.) with the collaboration of C. O. Sauer, M. Bates and L. Mumford: *Man's Role in Changing the Face of the Earth.* International symposium, Wenner-Gren Foundation for Anthropological Research, 1956.
Williamson, H.: *Salar the Salmon*, 1935.
Wimpenny, R. S.: *The Plankton of the Sea*, 1966.
Wollaston, H. J. Buchanan: *Plaice Egg Production*, 1926. (Min. of Agriculture and Fisheries.)
Worthington, E. B.: *Inland Waters of Africa*, 1930.
Young, J. Z.: *The Life of the Vertebrates*, 1950.

Index/names

Aarup, 141, 144
Albrecht, W. A., 46, 47
Apstein, C., 163
Arago, F., 161
Arrhenius, O., 31, 32
Atkinson, G. T., 163–4

Baerends, G. P. and Baerends van Roon, 131–4, 141
Balfour, Lady E., 49–51
Bell, A., 49, 51
Beresford, T., 52
Bertalanffy, L. von, 174
Bertin, L., 113–15
Beverton, R. J. H., 174–5
Black, J. D., 179, 192
Blackett, P. M. S., 174
Borley, J. O., 118
Braarud, T., 7
Brenchley, G. H., 53–6
Brierley, J. K., 208, 209, 214, 215
Butcher, R. W., 81
Buxton, P., 135

Carrighar, S., 136–7
Carruthers, J. N., 156
Cave, T., 53, 57
Clarke, G. L., 71, 80
Collingwood, Admiral, 85
Cooper, C. F., 94
Creear, T., 200
Crowcroft, P., 127–8, 134

Darling, F. F., 130
Davidson, D. F., 183
Debenham, F., 12
Dent, R., 147
Dobbs, C. M., 141
Dorst, J., 108, 112

Edwards, P. St. J., 209
Edwards, V. C. W., 131, 149
Elliot, R. H., 42–3, 47
Ellis, E. A., 11
Elton, C., 102, 159

Fell, H., 46, 51–4, 56
Fistein, B., 135

Forsyth, S., 195
Frankland, Dr., 190
Frost, S., 40, 132, 201, 203
Furneaux, B. S., 53–4

Galton, Sir F., 94
Gardiner, A., 173
Gardiner, R., 48
Gardiner, S., 151, 155, 156
Garstang, 118, 119
Graham, M., 4, 7, 29, 66, 72, 118–119, 129, 131, 134, 139, 143, 149, 154, 162, 173, 205, 209, 225
Graham, R. B., 195–7, 204
Gran, H. H., 7
Gulland, J. A., 166, 170, 174

Haeckel, E. H., 1, 2, 3, 168
Hall, Sir D., 48
Hall, I. G., 209
Hanson, C. O., 84, 90
Hardy, Sir A., 6, 104, 122, 177
Hasler, 112
Heape, A. S., 134
Hediger, H. P., 127
Hensen, V., 72, 162–3
Hjort, J., 170
Holt, S. J., 174–5
Hulme, H. R., 174
Huntsman, A. G., 116–17, 121, 128
Huxley, T. H., 164

Jahn, G., 170
Jamieson, Dr. A., 4
Jones, H., 49
Jones, N. S., 69–70
Jones, R. H., 112, 122

Kemp, S., 173
Kershaw, W., 1, 135
Kirkman, F. P., 130–1
Kluvers, H. N., 135
Kramer, G., 111–12
Kropotkin, P. A., 106

Laplace, M. de, 161, 164
Law, F., 205

Le Bon, G., 135
Lebour, M., 6
Lee, A. J., 7, 8
Leopold, A. S., 125–7, 192
Liebig, J. von, 46, 51
Lincoln, F. C., 164–5
Livingstone, D., 137
Lloyd George, D., 48
Lowe, J., 57

Macan, T. T., 64, 65
Machulich, 139
Malthus, T. R., 126
Manly, B. F. J., 166–7
Marais, E., 137
Mather, Sir W., 11
Meek, A., 115–16, 120–1
Mellor, W., 110
Meyricks, J., 57
Miller, E. C., 46
Mitcheson, J. C., 213
Morand, 161–3
Morrow, G., 129–30, 134, 158–9
Mortimer, 44
Mumford, L., 233
Murray, Sir J., 64, 174

Napoleon, 88, 161
Neal, E. G., 96, 98, 99, 101–3, 107, 121, 123, 124
Norman, J. R., 100, 115

Ottestadt, P., 170
Oxenham, J., 217
Oyler, P., 48–9, 108, 110

Parr, M. J., 105, 166–7, 190
Pearsall, W. H., 13, 33–4, 43, 62, 67
Pearson, 92–3, 188
Percy, R., 53–4
Petersen, C. G. J., 119, 163–5, 167, 168, 170, 171, 173
Popham, E. J., 5–6, 63, 64, 68
Portsmouth, Lord, 49
Pottenger, F. M., 46

Rasmussen, 171

Regan, C. T., 100
Richardson, J. A., 209
Robinson, D. H., 57, 59
Robinson, J., 214
Rowe, 127
Ruskin, J., 1, 13
Russell, E. J. and E. W., 13–14, 23, 31–3, 48, 99
Russell, E. S., 168–71, 173

Salisbury, E. J., 33
Sauer, C. O., 112
Savage, R. E., 6, 7
Savory, C. A. R., 191–2
Schmidt, J., 113–15
Selous, 122
Simpson, A. C., 163, 170
Southern, H. N., 156, 165
Southward, A. J., 68
Stapledon, Sir G., 13, 34, 42, 51, 53, 232–3
Steinbeck, J., 48, 188
Strowger, S., 137

Taaning, A. V., 3–4, 8
Taylor, A. J., 165
Thirgood, J. V., 208
Thompson, Sir D., 168
Thompson, W. F., 170, 176

Valentine, 87
Van Straelen, V., 189

Waller, R., 50
Watson, G. G., 183–4
Weir, J., 196
Wells, P., 135
Whitehouse, Cdr., 151
Williamson, H., 116
Wimpenny, R. S., 7, 77
Witcomb, D., 116
Wollaston, H. J. B., 162–3
Wood, R. F., 208
Worthing, E. B., 64, 65, 143–4, 155, 158, 183

Young, J. Z., 138–9, 167, 170

Index/places

Aberfan, 218
Aberystwyth, 31
Africa, 110, 122, 123, 129, 135, 141–160, 189
Antarctic, 108–9, 121, 176, 177
Arctic, 5, 108–9, 121, 138
Argentine, 109
Atherton, 214, 215

Baltic Sea, 116
Belfast Airport, 229–30
Bermuda, 114
Bickershaw, 209
Bickerstaffe, 208, 209
Bolton, 11, 87, 106, 224
   canal, 33
Borrowdale, 194–200, 203, 205

Canada, *see also* New Brunswick, Newfoundland, Nova Scotia, St. Lawrence River, 86, 88, 89, 108, 128, 136, 139, 164, 170
Clifton Park, 43
Coombe Beck, 197–200
Cumberland, 204
Cyprus, 183

Dale Fort, 166
Dogger Bank, 68, 72, 73, 119

Eccles, 210
Emin Pasha Gulf, 8, 152–5, 159
English Channel, 73
Esthwaite, 2, 67

Faroe Islands, 4, 8, 73
   Bank, 4
Forest of Bowland, 87, 205
Forest of Dean, 124

Grange-over-Sands *see* Merlewood
Greenland, 6, 108, 116

Haughley Farms, 50
Hindley, 210

Iceland, 8, 73, 108, 120, 121

Kavirondo Gulf, 146, 149, 153, 154
Kentmere, 180, 200–2, 206
Kentucky Experimental Farm, 47
Kersal Moor, 11, 93, 188, 190

Lake District, *see also* Borrowdale, Coombe Beck, Esthwaite, Kentmere, Newby Bridge, River Derwent, River Dunsop, River Kent, Rosthwaite, Wastwater, Windermere, 33, 43, 180–1, 194–206
Lincolnshire Wolds, 58
Long Island Sound, 82
Longton, 110
Lowestoft, 4, 7, 35, 137, 148, 155–6, 174, 195, 219

Madagascar, 184
Malta, 6
Manchester, 11
Merlewood, 90, 190
Minnesota, 204

Naples, 114
New Brunswick, 88, 116–17
Newby Bridge, 194–5
Newfoundland, 6, 8, 85
New Zealand, 24
Norfolk Broads, 11
North Sea, 2, 4, 8, 66, 68, 72, 73, 75, 104, 110, 118, 120, 121, 139, 140, 162, 163, 166, 168, 169, 171, 172, 173, 175
Norway, 8, 108, 116, 120, 121, 138, 139, 170, 178
Nottinghamshire, 57, 209
Nova Scotia, 88
Nova Zembla, 3, 6, 108

Paris, 161–2
Pennsylvania, 179, 207
Port St. Mary, 70, 128

Rhodesia, 191
River Derwent, 195–7
   Dunsop, 205

River (*cont.*)
  Irwell, 11
  Kent, 200, 202
  Lune, 164
  Maas, 73, 81
  Mississippi, 179, 192–3
  Rhine, 81
  Ribble, 68, 110
  St. Lawrence, 136
  Severn, 110, 113
  Thames, 73, 74, 81
  Trent, 81
  Tyne, 116
Rivington, 89, 105–6, 185–7, 194, 226
Romney Marsh, 29
Rosthwaite, 195
Rothamsted Experimental Station, 1, 9–10, 13, 15, 46, 48, 89, 194
Roudsea Wood, 90, 190

St. Helena, 184
St. Helens, 210, 215, 223
Salford, 12, 33, 60, 63, 86, 105, 165, 188, 190, 209, 215
  University, 1, 11, 166, 209, 213, 215, 217, 226, 231
Sargasso Sea, 114
Saskatoon, 39
Scotland, 64, 97, 145
Spitzbergen, 8, 78
Suffolk, *see also* Lowestoft, 49

Thurlbear Wood, 99–102, 121, 123

Victoria Nyanza, 141–60, 229
Virginia, 3

Wastwater, 67
Westhoughton, 213, 215
Westleigh, 215
Wigan, 106, 143, 223
Wiltshire, 57
Windermere, 65, 194

Index/subjects

Acidity, *see* pH
Agrostis, see Bents
Alder, 63, 87, 194, 196, 197, 199, 203, 204, 226
Algae, 14, 19, 37, 61, 69, 81, 180, 182, 208
  blue-green. 19, 36, 61. 180
Alkalinity *see* pH
Alsike, 216
Alternative husbandry, *see* Lea (ley) farming
Aluminium, 23
Ammonia, 21
Ammonium, 23–4
Amoeba, 18
Amphibia, 60–3
Angle of repose, 182
Anions, 23
Antibiotics, 21
Ants, 15, 112
Aphids, 97–103, 108, 121–5
Apple scab, 30
Arable husbandry, 41–4, 58
Arctic tern, 108–9, 121
Arsenic, 21
Ascaris, 18
Ash, 33, 83–5, 90, 91, 93, 94, 97, 98, 194, 198, 200–4
Autecology, definition and examples, 2–6, 9, 10, 168, 205
Autocatalytic theory, *see* Sigmoid theory
Auxospores, 67

Baboons, 137–8
Bacteria, 17–24, 31, 33, 36, 37, 40–43, 47, 53, 56–7, 77, 80–2, 89, 102–3
  actinomycetes, 20, 50, 56
  azotobacter group, 36
Badgers, 98, 101–4, 124
Barley, 52, 58
  mildew, 55
Basic slag, 34
Bats, 101, 103, 104
Beavers, 85–6, 205
Beech, 87, 90, 91, 97
Bees, 112

Beetles, 15, 16, 28, 39, 60, 98, 103, 104, 189
Benthos, 71
Bents, 30, 31, 182, 188, 216
Birch, 33, 86–8, 91, 97, 208
Birch Coppice Colliery, 213
Botany Bay plot, 210
Bristle worms, 18
British Ecological Association, 12
Broom, 36
Burnet, 35

Cadaverine, 22
Canopy, 84, 96–9, 107, 122, 126, 190, 229
Carbon dioxide, 18, 19, 32, 38, 41, 46, 77, 180, 183, 209
Carbonic acid, 35, 90, 180, 183
Carlsburg Foundation, 113
Caterpillars, 98, 101, 103, 182
Cations, 23
Catstail, 30
Cattle, grazing and farming, 15, 16, 25–7, 40–1, 49, 51, 52, 57, 90, 135, 187, 191, 196–7, 203, 230
Cedars of Lebanon, 183–4
Centipede, *see* Myriapod
Chalk, 87, 182
Challenger Society of Oceanography, 173
Chemics plot, 210
Chicory, 35, 216
Chiff-chaff, 100, 109
Children, 219–27
Chlorophyll, 18, 68–9
Cichlids, 131–3, 141
Civic Trust, 224
Cladocera, 61, 80
Clay, 21–3, 37–8
Click beetle larvae, *see* Wireworms
Clifton Moss plot, 210, 214
Climax vegetation, 86–7
Clover, 31, 35, 41–2, 53, 188, 215, 216, 225
Cocksfoot, 31, 60, 188, 208, 216
Cod, 2–4, 6–9, 66, 78, 120–1, 124
Community Council of Lancashire, 209, 217, 221

Contour ridging, 189, 208, 210, 214, 215
Copepoda, 61, 67, 72, 80
Coral fish, 112
Coregonids, 67
Coriolis force, 78–80, 228
Corn, 41, 52, 55, 57
Corsican pine, 91
Coypu, 11, 229
Crustacea, 71, 80
Currents, 7, 63–8, 79, 80, 117, 121, 205, 228
Cyprinids, 64

Dandelion, 35
DDT, 233
Deer, 90, 126–7, 190
Determinants,
  examples, 7, 11, 59, 62, 63, 68, 73, 74, 77, 80, 95, 96, 107, 112, 121, 122, 156, 228–32
  theory, 10, 228
Detritus, 62
Diatoms, 6, 19, 61, 67, 73, 80, 152
Difflugia, 18
Dragonflies, 104, 166
Drift bottles, 147–8, 151, 156
Dung, 26, 29, 39–41, 45–8, 50, 51, 56, 103, 104
Dungflies, 104
Dustbowls, 48, 57, 58, 188–9, 192

Earthworms, 15, 16, 21, 22, 24, 28–31, 33, 36, 39, 90, 99, 102–4
Ecology
  origin of term, 1, 209
  University course, 1, 11, 232
  examples, 2, 9, 10, 90, 100, 107, 108, 152, 229–31
  farm, 13–59
  human, 224–7
Eco-systems, 8, 10, 39, 59, 68, 71, 80, 107, 108, 122, 124–5, 228–231
Eels, 111–13, 117, 120
Eel-worm, *see* Nematodes
Elephant grass, 189–90, 212
Elm, 83, 91
Elvers, 113–14
Enzymes, 37
Epilimnion, 66
Erosion, 178–218, 229
  by animals, 183–5, 189–92
  by people, 185–8
  by water, 194–206
  by wind, 188–9
Eskimos, 7, 171

Estuaries, 68, 228
Euglena, 18

Farmers' Club, 51, 57, 192
  Journal of, 51, 53, 82
Farming to leave, 192
Ferns, 180, 200, 204
Ferric ion, 32–3
Ferrous ion, 32–3, 43
Fertility cycle, 39–41, 80
Fertilizers, 24, 45–53
Fir, 88, 90, 97
Fire, 88–9, 94
Fishing, 72–7, 141–60, 166–77
Flagellates, 6, 17, 18
Flycatchers, 125
Fog grass, 216
Forestry Commission, 94, 207
Forests, 83–9, 228
Four course shift, 42, 52, 54
Four-gates plot, 210, 213
Foxes, 98, 101–4, 124, 190
Friends' Work-camps, 214, 215, 217, 224, 226
Frugal reclamation, 210, 215, 217, 218
Fulvic acid, 23
Fungicides, 55
Fungus, 16–20, 26, 36–8, 50, 55–7, 81, 87, 98, 99, 102–4, 197

Gannets, 8
Glass-eels, *see* Elvers
Glyceria aquatica, 33, 60, 61
Goats, 25, 183–5, 191
Golden plover, 109
Grass, 25–43, 49–50, 207, 208, 215–216
Grazing and browsing, 10, 25, 40, 131, 134, 191, 192, 202, 203, 229, 230
Grilse, 117
Ground nuts scheme, 88
Gulbenkian Foundation, 217, 223
Gulls, 164

Habitats, 2, 6, 11, 96, 98–101, 126, 150
Haddock, 5, 139
Hair grass, 30, 188, 208, 216
Halibut, 170, 176
Hares, 138–40, 222, 229–30
Hay, 26, 36, 41, 42, 204, 210, 215
Hedgehogs, 104
Hedges, 57, 58, 192, 195, 202–3
Herring, 4–9, 73, 81, 120, 121, 124, 137–9
Hornbeam, 91

Horses, 25–6
House martins, 105
Humic acid, 23
Humin, 23
Humus, 22–3, 30, 32, 35, 38–42, 46–52, 56, 57, 59, 90, 99, 180, 182, 183, 190, 192, 228, 229
  alpha, 23
  beta, 23
Hydrogen-ion, 32
Hypolimnion, 66–7
Hypsometer, 92–3

Id-gate plot, 210
Impala antelope, 191
International Council for the Exploration of the Sea, 168, 174
Inter-tidal zone, 68–70
Iron oxide, 209

Journal of Ecology, 33

Kaie, 53, 57
Kavirondo tribe, 143–59
Kingfishers, 60, 80, 128
Kinship of Husbandry, 48–9
Krill, 8

Ladang, 87–8
Ladybirds, 101, 102, 124
Lakes, see also Victoria Nyanza, 64–8, 228, 229
Larch, 90, 91
Lea (ley) farming, 42, 45, 47, 49, 51–4, 56, 182
Lemmings, 134, 138
Leptocephali, 113–14
Ley farming, see Lea (ley) farming
Lichens, 14, 37, 98, 105, 180
Lignin, 17
Lime, 21, 30, 32–5, 56, 69, 87, 182, 188, 209, 210, 215, 216
Limestone, 87, 107, 182, 209
Limpets, 69–70
Lincoln Index, 164, 166
Lochs, 64
Lucerne grass, 31, 35
Lupin, 36
Lynx, 138–9

Macro-ecology, definition and examples, 7, 9, 10, 45–59, 63, 73, 173, 176, 177, 189, 203, 205, 228
Maize, 88
Manganese dioxide, 32
Manganic ions, 32

Maple, 88
Marking experiments, 116–20, 163–164, 166–7, 175
Merganzers, 128
Mesh regulations, 155, 158–9, 169, 170, 174
Mesta, 185
Mice, 98, 127–8, 131, 134, 165
Migration, 2, 3, 108–23, 134–5, 165, 229
Migration fidgets, 111
Milk, 52
Millipede, see Myriapod
Milt, 115
Mites, 15–17, 19, 28, 36–7, 56, 99, 102, 104
Moles, 15, 28, 90, 104
Molinia, 34
Moose, 136–7, 140
Mor, 90, 99, 102
Mormyrus, 150
Moss, 14, 37, 86, 180, 198, 199, 208, 216
Moths, 101, 166–7
Moulds, 17
Mud, 22, 61–3, 68, 151, 228
Mull, 30, 90, 99, 102
Mussels, 82, 177
Mycelium, 104
Mycorrhiza, 16, 20, 34
Myriapods, 15, 28, 99
Myxosporidia, 19

Nematodes, 17, 18, 36–7, 56
Newton Road plot, 210
Ngege, 141–59, 171
Niches, 84, 90, 99–101, 121, 126
Nightjar, 101
Nile perch 158
Nitrates, 9, 21, 23, 24, 32–4, 36, 42–6, 77, 229
Nitrite, 21, 32, 43
Nitrogen, 9, 19–21, 23–4, 36, 40–4, 53
North Atlantic Drift, 78, 114–15, 120
North West Economic Planning Council, 11

Oak, 33, 46, 83–5, 87, 89–91, 93, 94, 97–8, 103, 105, 106, 187, 203, 204
Oceans, 75–80, 228
Osiers, 87
Ostracoda, 61
Owls, 101, 103, 104, 124
Oxygen, 14, 20, 21, 22, 32, 38, 44, 67

Papyrus, 142, 143, 147, 151–3
Parasites, 16, 18, 19, 21, 41, 42, 56, 57, 103, 135, 136
Parr, 116
Peat, 31, 35, 40
Pecking order, 131, 134
Penguins, 77
Periphyton, 61, 62, 150–3
pH, 11, 30–6, 45, 62, 69, 81, 87, 90, 97, 157, 188, 215
Phages, 21, 24, 37
Phosphates, 8, 34, 46, 67, 72–4, 77, 81, 216
Phosphorus, 23–4, 34
Photosynthesis, 66–7, 81
Phytoplankton, 9, 61, 71, 72, 101
Pike, 61–2, 101, 124, 137
Pine, 90, 97
Plaice, 74, 75, 118–21, 159, 162–3, 175
Plankton, 6–8, 72, 77, 108–9, 113, 114, 117, 120, 121, 162–4, 229
Plantain, 35
Plant succession, 11, 86, 87, 180, 182, 208, 209, 216
Pleurococcus, 19, 98, 180, 182, 208
Polar bears, 7
Pollution, 81–2, 105–6
Ponderoso pine, 94
Ponds, 60–4, 228
Pondweed, 61–3
Poplar, 87, 91, 218, 224
Populations, 100, 101, 123, 128, 138–140, 161–77, 229
Potassium, 23–4, 46
Potato, 56
  blight, 55
Predation, 6, 18, 21, 67, 69, 76, 101–103, 124–31, 134–7, 158, 167–169, 192, 229, 231
Pretoria tip, 214, 215
Protein, 20, 21
Protopterus, 151–2
Protozoa, 18, 19, 21, 36–7
Putrescine, 22

Quadrat method, 165

Rabbits, 10, 25, 101, 102, 190, 191
Race 60, 55
Radiation, 82
Rape Grass, 225
Redds, 115
Redox, 32–5, 43, 62, 68, 69, 81
Reindeer, 125, 136, 164
Rhinoceros, 191
Rhizosphere, 22

Rice, 12, 22, 43–5, 62
Roach, 63, 165
Robin Hood plot, 210
Robinia, 218
Robins, 130–1
Root crops, 41, 52, 53, 57
Roots, 14, 15, 27, 29, 34–9, 41, 56, 84, 90, 98, 183, 190, 191, 194, 198, 200, 203, 204, 213, 216, 217, 228
Rothwell perdix, 55
Roundworm, *see* Nematodes
Rowan, 200, 203
Rucks, 222
Rye grass, 31, 216, 225

Sainfoin, 31
St. George's plot, 210
Sallows, 34, 65, 87, 204, 210, 212, 214, 215
Salmon, 115–17, 121, 128, 164
Salmonids, 64, 67
Saprophytes, 19, 104
Scholasticism, 231–3
Seals, 7. 8
Seashore, 68–70
Seaweed, 68–70
Shale, 86, 190, 207–9, 215, 216, 226
Sharks, 8
Sheep, 25, 26, 30, 34, 52, 57–9, 105, 106, 135–6, 186–7, 191, 196–199, 203, 205, 206, 228
Sigmoid theory, 170
Silicon, 23
  tetrachloride, 23
Slag, 207, 215, 223
Sleeping sickness, 135, 146, 153
Smolt, 116
Sodium silicate, 23, 157
Soil, *see also* Subsoil, 178–9, 190, 192
  acidity and alkalinity, 11, 30–4
  formation, 180, 182
  husbandry, 48, 90
  lattice, 14, 22, 23, 38
  science, 2–44
  solution, 14, 24, 32, 34, 60, 61
  types
    bog, 34
    moorland, 34
    woodland, 33
Soil Association, 49
Sole, 2, 168
South east trade wind, 148, 151, 156, 229
Sparrowhawks, 101–4, 124
Spawning, 2, 113–15, 119–21
Spiders, 15, 17, 104

Spoil heaps, 86–7, 182, 190, 207–27, 229
Springtails, 16, 17, 36, 99, 102, 104
Spruce, 90, 91, 97
Squirrels, 85, 98, 101, 103
Stain, 26, 42, 134
Starlings, 28, 29, 111–12, 124, 229
Stoney lane plot, 210
Streams, 63–4, 228
Subsoil, 23, 34, 90, 107, 180, 217
Sulphate, 32
Sulphite, 32
Sulphur, 21, 105–6, 121, 209
Swallows, 109–10
Sycamore, 63, 86–7, 90, 91, 93, 94, 194, 197, 199, 204, 205
Synecology, definition and examples, 5–10, 25–45, 54, 56, 60, 71, 124, 125, 168, 189, 205
Synodontis, 150, 157–8

Taenia multiceps, 135
Take-all, 55
Tapeworm, *see Taenia multiceps*
Terraces, *see* Contour ridging
Territory, 130–4
Thermocline, 65–7, 73, 74, 77, 82, 228
Thompson's gazelles, 129–30
Tilapia esculenta, *see* Ngege
Timber
  properties, 83–5, 91
  volume, 91–5

Timothy grass, 30, 31
Tits, 86, 95–101, 121–2, 124, 134
Trace elements, 24, 34
Trypanosome, 135
Turbot, 2

Unicorns, 129–31, 134
United States Conservation Service, 179, 192

Vandalism, 186–7, 224–7, 229

Warblers, 86, 100, 108, 109, 111, 112, 122, 123
Wenner-Gren Symposium, 94
Whales, 8, 77, 81, 170, 171, 174, 176–7
Wheat, 39, 55, 58, 228
White bass, 112
Willow, 91, 204, 213, 218
Wireworms, 15, 24, 27–9, 36, 50, 56, 124
Wolves, 125, 126, 136–7
Woodlice, 99, 103, 104
Woodpeckers, 98, 104
Worm-casts, 30

Yeasts, 20, 171
Yellow pine, 86, 87
Yellow rust, 55
Yew, 87

Zooplankton, 61

SOMERVILLE COLLEGE
LIBRARY